Advances in Biology

This is a numbered series focused on the latest research in biology.

nova
science publishers
www.novapublishers.com

Advances in Biology

Advances in Biology. Volume 4
Charles D. Grant (Editor)
2023. ISBN: 979-8-88697-999-2 (Hardcover)
2023. ISBN: 979-8-89113-074-6 (eBook)

Advances in Biology. Volume 3
Charles D. Grant (Editor)
2023. ISBN: 979-8-88697-779-0 (Hardcover)
2023. ISBN: 979-8-88697-846-9 (eBook)

Advances in Biology. Volume 2
Charles D. Grant (Editor)
2022. ISBN: 979-8-88697-422-5 (Hardcover)
2022. ISBN: 978-1-61761-293-0 (eBook)

Advances in Biology. Volume 1
Charles D. Grant (Editor)
2022. ISBN: 979-8-88697-201-6 (Hardcover)
2022. ISBN: 979-8-88697-218-4 (eBook)

More information about this series can be found at
https://novapublishers.com/product-category/series/advances-in-biology/

Charles D. Grant
Editor

Advances in Biology

Volume 4

nova
science publishers

www.novapublishers.com

Library of Congress Cataloging-in-Publication Data

ISBN: 979-8-88697-999-2
ISSN: 2833-1753

Published by Nova Science Publishers, Inc. † New York

Contents

Preface

This volume contains twelve chapters that detail advances in biology. Chapter One discusses non-surgical treatment for periodontal diseases. Chapter Two develops a comprehensive understanding of the influence of metal pollution on ecosystem disturbance. In Chapter Three transcriptome analyses were conducted using the cyanobacterium Synechococcus elongatus PCC7942 exposed to severe salinity and antibiotic stresses. Chapter Four presents a description of plant lectins, in relation to structure, localization, heterologous expression, detection and purification methods, biological activity, and biotechnological applications. Chapter Five discusses the use of methods of multivariate statistics in order to identify significant differences between the histochemical activity of enzymes of the skin glands of laboratory and wild rats. Chapter Six proposes fucoidan as a promising nutraceutical to improve gastrointestinal function. Chapter Seven discusses functions of α-SMA in HSC activation and its potential as a therapeutic target for antifibrotic therapy. Chapter Eight describes the audio vestibular disorders in children with Cytomegalovirus. Chapter Nine explains loop-mediated isothermal amplification. Chapter Ten studies the effect of short-term liquid nitrogen storage of seeds on the germination, phenotypic characteristics, phenolics compounds, soluble sugar and protein contents in *L. leucocephala*. Chapter Eleven reviews the biological properties of Sphingosine-1-phosphate (S1P) and summarizes the current knowledge of the function of S1P in macrophage biology. Lastly, Chapter Twelve explains the structure of the neutrino halo of a chicken egg.

Chapter 1

Non-surgical Periodontal Treatment II: Scaling and Root Planing

Luis V. Maita Véliz*
L. Mariano Maita Castañeda
and Maria Castañeda Mosto
Faculty of Dentistry, National University of San Marcos, Lima, Peru

Abstract

Periodontal diseases are one of the most prevalent in the world population.

Its etiology is multifactorial, but periodontal pathogenic microorganisms, known as biofilm, play an important role in its development. Its effects are negative and cause tooth loss. In addition, at a systemic level they can cause alterations in other organs, such as the heart or lungs, including premature births.

A timely diagnosis and adequate treatment are ideal procedures for its prevention and elimination.

Clinical research in recent decades has demonstrated the efficacy of non-surgical periodontal procedures for the treatment of periodontal disease.

Periodontal scaling and root planing techniques remain the "gold standard" for the treatment of periodontal disease.

Both procedures correctly performed are an essential part of the treatment of these diseases, even when periodontitis is in very advanced stages, allowing full recovery of periodontal health.

---

* Corresponding Author's Email: lmaitav@unmsm.edu.pe.

In: Advances in Biology. Volume 4
Editor: Charles D. Grant
ISBN: 979-8-88697-999-2

Dentists in general should seek to access better academic training on these non-surgical periodontal procedures to prevent millions of people in various parts of the world from suffering from periodontal diseases and losing their natural access to a good quality of life.

Keywords: periodontal diseases, scaling, root planing

Introduction

Biofilm is considered the main factor in the etiopathogenesis of periodontal diseases.

The basic procedures to combat periodontal infections continue to be the removal of supragingival and subgingival bacterial deposits by means of scaling and root planing.

Non-surgical mechanical treatment is the cornerstone of periodontal therapy and the main recommended method of controlling periodontal infections (Cobb, 1996).

Periodontal diseases are one of the most prevalent in humans. It affects not only the loss of teeth, but is also related to systemic diseases, altering the functioning of organs such as the heart, lungs, reproductive system and others (Maita et al., 2003).

The cost of care for treatment in industrialized countries is very high. Periodontal treatments of affected patients reach millions of euros in European countries and in other parts of the world (European Federation of Periodontology, 2021).

The main factor in the etiopathogenesis of periodontal diseases is the biofilm (bacterial plaque). To this are added other factors: environmental, genetic, periodontal hygiene, individual tissue response (of each person), periodontal control and maintenance (Page et al., 1997; Page et al., 1997).

The phase to start periodontal treatment is related to removing bacterial plaque or biofilm and dental calculus at the supragingival and/or subgingival level. The use of antimicrobial agents as a complement may be necessary.

In many cases, non-surgical therapy alone is enough to restore periodontal health (Newman et al., 2018).

Initial periodontal therapy has been described as the approach directed at the removal of biofilms, toxins, and dental calculus. And the restoration of a biologically acceptable root surface (Dentino et al., 2013).

All periodontal treatment aims at the meticulous removal of bacterial deposits at the supragingival and subgingival level. Due to the anatomical architecture of the gingival and osseous soft tissues, they require very detailed observation clinically to achieve this goal.

An example is the clinical removal of cement from the periodontal pocket. Mechanical procedures must be somewhat conservative, not very aggressive because the dentin can be exposed and cause a complication.

Clinical changes in inflammation are good indicators of the progress of periodontal treatment (Adriaens et al., 2004).

Before starting periodontal treatment in patients, the patient should be carefully evaluated. This evaluation must be permanent until the maintenance phase, only then can success in periodontal treatment be considered (Maita et al., 2022).

Clinical investigations in the last three decades have demonstrated and confirmed the proven efficiency of non-surgical periodontal procedures for the treatment of periodontal diseases. This is why these periodontal procedures are still called the "gold standard" of periodontal therapy (Badersten et al., 1981; Badersten et al., 1987; Maita et al., 2016).

In recent years, these periodontal procedures have been improved by new technologies in instrument design, especially for somewhat inaccessible areas such as furcations, interproximal areas, distal surfaces of upper molars, root grooves, and deep pockets, among others.

These periodontal instruments designed for these almost inaccessible areas now allow for better dental calculus and biofilm removal. In the future they will probably further improve the design of these periodontal hand instruments (Oda et al., 2004).

Although it is true that manual mechanical debridement allows the elimination and disorganization of the periodontal pathogenic flora, this could be complemented with the use of antimicrobial agents in order to eliminate the remnants of residual pathogenic agents.

Usually, these periodontal procedures can be performed by quadrants or sextants. Some research-based clinicians employ the "full-mouth disinfection" technique, in 24 hours and supplementing with the use of chlorhexidine for the probable bacterial reservoirs.

In all treated cases we must carry out a reassessment after about 3 months to 9 months (Badersten et al.,1984).

This reassessment makes it possible to observe the results obtained in order to recommend whether further complementary treatment is necessary, which could be surgical treatment. Due to the chronic nature of periodontal diseases and in order to recover periodontal health, it is a periodontal principle to carry out the periodontal maintenance period or phase. This will be done according to the needs of each patient, and it cannot be generalized to all with the same standard pattern of periodontal maintenance for the intervals of office visits.

The prevalence of periodontal diseases (chronic periodontitis) varies between 30% and 60% in adults. In children, this prevalence ranges from 30% to 80%, reaching an average of 35%, which we can consider to be of concern because a percentage of this population will suffer from chronic periodontitis in adulthood (Botero et al., 2015).

These youthful populations at risk for periodontal disease should be better served by the corresponding health institutions: national health services, private health companies, and private country offices, putting into practice promotion, prevention, and education strategies at the national level, individual and population, in people with periodontal diseases and especially in healthy people (Contreras, 2016).

Microbiological Aspects in Relation to Scaling and Root Planing

Research to date indicates that periodontal diseases have a microbial origin. Its particularity is that not a single microorganism intervenes, but a group of microorganisms.

Costerton defines a biofilm as a bacterial community in a liquid medium, characterized by bacteria that are attached, either to a substrate or surface, or to each other. These bacteria are located in an extracellular matrix produced by themselves and show an altered phenotype in terms of cell multiplication or the expression of their genes (Donlan et al., 2022).

To eliminate biofilm, it is essential to break its microbial structure, which can be achieved by mechanical procedures, such as: technically appropriate brushing, flossing, professional prophylaxis, scaling, root planing, surface polishing, dental restorations with a seal able to prevent the development of biofilm.

Since the first bacterial description, carried out in 1683 by Anthony van Leewenhoek, outstanding investigations over time, but even more so in recent decades, have allowed a better understanding of the biofilm process.

A little more than 500 bacterial species have been found with the capacity to colonize subgingival plaque, while there are many other bacteria that are not cultivable. We do not know the function of the latter and they could well be related to the onset and/or development of periodontal diseases (Moore et al., 1982).

Until now, it is not known exactly which is the microorganism that initiates periodontal disease, or which are the periodontal pathogens that participate in it, which would give us more options and better strategies for the treatment of periodontal diseases.

Periodontal pathogenic bacteria need to adhere to periodontal tissues for the initiation and development of periodontal diseases. One of the mechanisms of adhesion to the oral epithelium is the use of its bacterial fimbriae, especially the gingival fimbriae of Porphyromonas (Hamada et al., 1998).

In cases of chronic periodontitis (Contreras et al., 1998) and juvenile periodontitis (Ting et al., 2000), it has also been possible to detect some types of viruses, such as herpes simplex, cytomegalovirus, and Epstein-Barr virus type 1, in the crevicular areas of patients.

When cells are infected by the herpes virus, they can decrease the host's defense mechanisms, causing periodontopathogenic microorganisms to increase their population in the subgingival areas and invade surrounding periodontal cells and tissues.

The microorganisms that inhabit the supragingival plaque basically feed on saliva and food debris. Subgingival plaque microorganisms produce organic acids at the expense of carbohydrates. This subgingival plaque, when it matures, releases pro-inflammatory substances such as: lipopolysaccharides and peptidoglycans.

In the subgingival areas, there are periodontal pathogenic microorganisms, which, the more they are fed, increase in number and produce various types of exotoxins and enzymes, which will cause tissue damage and periodontal destruction (Maesh, 1994).

The presence of supragingival plaque is fundamental or a requirement for the formation of subgingival plaque.

Subgingival plaque is classified into three classes: Adhered, free or non-adhered, and epithelium-associated (Newman et al., 1984).

Adhered subgingival plaque has a direct contact relationship with the tooth surface. Its composition consists mainly of grampositive bacilli and cocci (Listgarten, 1976). They can survive in conditions of limited nutrition and in strictly anaerobic situations. They are considered stable in subgingival plaque. Subgingival plaque is estimated together with subgingival calculus and root caries.

The free or non-adhered plaque is formed by motile and gram- negative bacilli. It is made up especially of the periodontal pathogens, *P. gingivalis* and *T. denticola*. Their joint action generates an accelerated inflammatory process.

Epithelium-associated plaque is loosely adherent to the gingival epithelium and is made up of motile, gram- negative rods. An important microorganism in this plaque is Prevotella intermedia. Plaque associated with the gingival epithelium could play a role in periodontal pathogenesis and be involved in bacterial invasion of connective tissue (Carranza et al., 1983).

The living conditions of biofilm microorganisms can vary due to environmental factors. For example, the optimum pH for the growth of *P. gingivalis* is between 7.0 and 8.0. Another environmental factor is that there is inflammatory fluid and blood from the gingival crevice in the medium, its encounter with *P. gingivalis* makes it more virulent. What we want to point out is that the virulence of these microorganisms can vary according to environmental factors.

When periodontal scaling and root planing procedures are going to be carried out, these microbiological considerations must be taken into account because it will allow us a more precise mechanical elimination of periodontopathogens, alter their environment and the products that nourish them, finally achieving that the patient can recover their periodontal health.

The exhaustive knowledge about the microbiological nature of the biofilm is justified, since otherwise scaling and root planing would only be empirical techniques against periodontal pathogens, a group of highly organized microorganisms.

Not only are they responsible for the development of biofilm, but they also have a great capacity to colonize dental surfaces and reach very aggressive virulence indices.

Philosophy of Periodontal Treatment

After a thorough periodontal clinical examination and the correct periodontal diagnosis, treatment of periodontal lesions is carried out.

Traditionally, periodontal treatment consists of the following stages:

- Periodontal hygiene
- Scaling and root planing

- Periodontal reassessment
- Periodontal surgery
- Periodontal maintenance

Periodontal treatment always begins with periodontal hygiene procedures to remove bacterial plaque, located above or below the marginal gingiva. If it is not done properly, an inflammatory process will be installed in the gingival tissues, producing gingivitis.

Over time, plaque that is not removed will calcify and become dental calculus that adheres firmly to the tooth surface. The inflammation will persist and can reach the periodontal ligament and bone tissue, causing destruction of these tissues. This periodontal pathology is known as periodontitis. As you can understand, this phase is crucial for the recovery of periodontal health.

The scaling and root planing phase in many cases of periodontitis is the only definitive periodontal treatment to restore periodontal health.

Scaling and root planing is the "gold standard" of periodontal treatment in all periodontal diseases.

We agree with researchers, who state that it is the most important, fundamental and irreplaceable clinical procedure in periodontal treatment (Barletta, 2009).

They are the most important clinical procedures in periodontal treatment because they will reduce the bacterial load and alter the bacterial composition, managing to change the bacterial flora into one that is favorable for periodontal health. These microbiological changes reduce inflammation and provide relative stability of the periodontal attachment level (Petersilka et al., 2002).

Its importance also lies in the maintenance of periodontal health.

To develop scaling and root planing procedures, more time should be devoted to learning and mastering the fine art of manual scaling and root planing, which will allow excellent dexterity to be developed and are more effective and better mechanical debridements during periodontal surgeries (Grover, 2022).

Preparation, experience, skill, dedication, and professional skill are important elements in dental calculus removal (Brager et al., 1989).

Mechanical debridement must be accompanied by oral hygiene procedures to obtain periodontal health recovery of the affected periodontal tissues (Badersten et al., 1984).

In the gingival tissues, pathological infiltrative cells slowly disappear and neocapillaries and fibroblasts appear in their place. By leaving the dental surface smooth and free of irregularities and infection, an area biologically suitable for tissue healing will be produced (Adriaens et al., 2004).

It is important to point out in periodontal health that another clinical sign to consider is gingival bleeding on probing. Even bleeding on probing can be seen in apparently and clinically "healthy" gingival tissues and in very young populations. Bleeding is an objective, visible and reliable sign to detect periodontal alterations (Maita, 1992).

In periodontal treatments based on scaling and root planing, bleeding on probing decreases by up to 74% after 60 days (Badersten et al., 1981).

Gingival Groove

The gingival groove, gingival, crevicular or sulcular cleft, is a virtual cavity that surrounds the dental neck like a ring or necklace, is V-shaped and determines the cervical limit of the clinical crown of the teeth. It has a maximum depth of one or two millimeters and is limited on the inside by dental enamel, on the outside by the free or marginal gingiva (it is also called the soft wall of the sulcus), and on its apical part by the epithelium of the sulcus. Union (Phalotic et al., 2006).

It has a temperature of 36 degrees Celsius, a slightly alkaline pH which facilitates the development of a range of parasitic bacteria.

The gingival sulcus constitutes a microsystem capable of attracting multiple bacterial genera, especially strict and facultative anaerobics, due to its low oxidation-reduction potential, low oxygenation, and large amount of nutrients that are very attractive to bacteria. In this space is the gingival or crevicular fluid (blood filtered through the junctional epithelium). Said fluid contains enzymes, antibodies, complement factors, interleukins, and defensive cells, among other antibacterial factors (Phalotic et al., 2006).

The gingival sulcus has the particularity of being poor in oxygen, but with a large amount of protein. Therefore, it is capable of harboring and preserving large numbers of strict and facultative anaerobic bacteria. It is recognized as the most complete anaerobic niche that exists in the body (Farias, 2003).

Under normal conditions, a large number of bacteria inhabit the gingival sulcus. Gram positive cocci (*Streptococcus sanguis*, *S. oralis*), fusobacterium, prevotella, strict anaerobic bacilli, capnocytophaga, selenomonas, treponemas predominate. It is also a propitious field for elements such as porphyromona gingivalis, actinobacillus, actinomycetemcomitans, which invade cells and multiply within them. This bacterial infestation produces toxins and lytic enzymes (Farias, 2003).

The so-called "healthy gingival sulcus" is a point of contention. It is clinically healthy if the gingiva shows a normal coral-pink color, with stippling, probing depth less than 2 mm, and no bleeding on probing. From the microbiological or immunological point of view, we can almost always detect some degree of inflammation in the sulcus and an increase in gingival fluid, leukocytes and interleukins. It should be noted that there should not be any symptoms (Liebana, 2002).

A furrow with a total absence of defensive elements is found only in gnotobiotic animals, that is, free of germs, so a furrow with a small number of defensive cells is considered normal (Jhonson et al., 1995).

Dental Calculus

Dental calculus is any calcified deposit that forms on teeth and dentures. It is composed of inorganic (70 to 90%) and organic elements.

Recognizing the importance of dental calculus in the progression of periodontal disease and in order to successfully complete its treatment, entails a thorough understanding of the following aspects:

- The removal process in professional dental hygienic practice.
- Oral hygiene time invested in removing dental calculus and preventing its formation.
- The need for total removal of the dental calculus, even from areas of difficult access, because it allows the perpetuity of periodontal disease and/or the acceleration of its dental calculus development. What is considered a modifiable factor and is associated with the recovery of periodontal health.

The dental calculus acts as a reservoir for biofilm bacteria. Colonies of bacteria nested within the dental calculus are impossible to remove using known oral hygiene procedures; and rather provide "protected areas" in periodontal pockets, allowing the presence of bacteria in close proximity to the periodontal tissues (Perry et al., 2014).

Dental Calculus Types

According to its location with respect to the gingival margin, dental calculus is classified as supragingival and subgingival.

Supragingival Calculus

It appears above the gingival margin. It is yellowish white in color and clayey in consistency. It is installed near the exit orifices of the major salivary glands, also on the lingual surfaces of the lower anterior teeth and buccal surfaces of the upper molars.

It is deposited on the smooth and uniform surface of the enamel by superficial adherence, unlike subgingival calculus. (Figure 1)

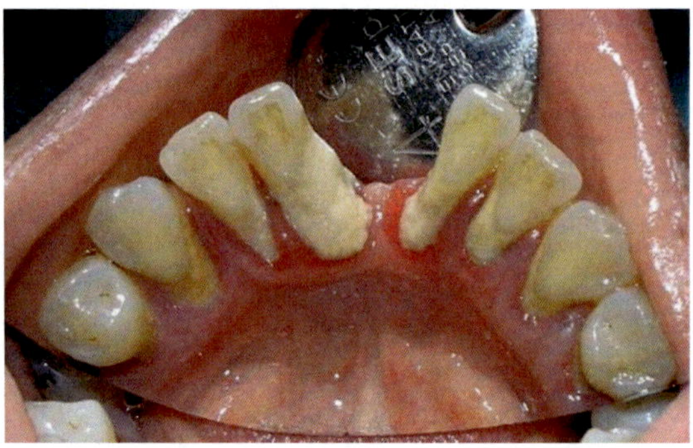

Figure 1. The patient presents copious amounts of supragingival calculus in the lower anterior area. Separation of the central incisor teeth can be seen.

Subgingival Calculus

It forms on root surfaces below the gingival margin and can reach periodontal pockets. It is of a different color than supragingival, ranging from dark green to black, probably because the plaque biofilm contains blood products associated with gingival bleeding. It is difficult to detect subgingival calculus. It is usually hidden and not directly displayed. The use of specific instruments is essential: periodontal probe, explorer and others. And something very important, locating this type of calculation requires the operator's trained skill, only his perseverance will improve the detection (Perry et al., 2014).

It is very laborious to remove it because unlike supragingival calculus, subgingival calculus has a special adhesion of the pellicle to the cement. Precisely its adhesion mechanism makes the difference with the other dental calculus. Also, the irregularities of the cement, where it is embedded, play an important role in its adhesion. Unlike supragingival calculus, it can appear anywhere in the oral cavity. Its distribution is more homogeneous (Perry et al., 2014).

Periodontal pockets almost always contain subgingival calculus, sometimes in microscopic amounts. The stone provides a reservoir for bacteria and endotoxins.

During the process of calcification and mineralization of the bacterial plaque, it can reach 90% in just 12 days (Figure 2).

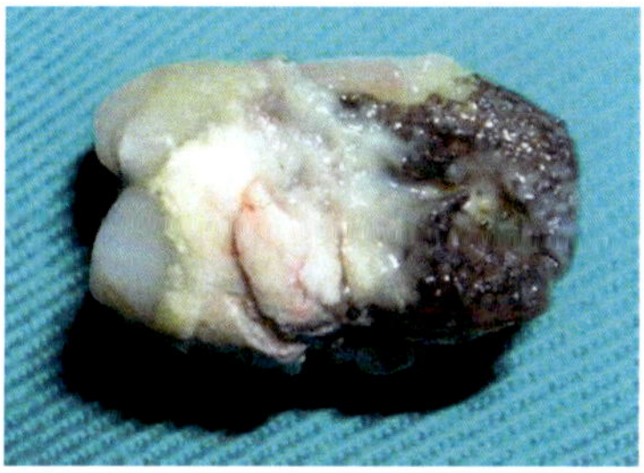

Figure 2. Molar extracted due to advanced periodontal disease. Abundant yellowish-white supragingival calculus and dark green and black subgingival calculus reaching the apex are observed.

Periodontal Probe

Periodontal probing is part of every patient's clinical dental examination to understand the clinical signs of health and periodontal disease. If the periodontal clinical examination is incomplete, the diagnosis will be incorrect and therefore periodontal treatment a failure. Visual modeling for performance just in time is another central instructional approach for effective mobile learning. This instructional strategy often involves reducing extraneous cognitive load on the spot, providing worked examples right before performing a critical task, and visualizing learning object in three dimensional presentations.

Using the Probe

The following technique is very useful for detecting periodontal pockets.

1. The probe is inserted below the marginal gingiva, and gently moved toward the junctional epithelium. Following the longitudinal axis of the dental piece.
2. It can be found during the passage of the periodontal probe, obstructed by a hard bed that does not yield: dental calculus.
3. The probe is gently lifted away from the tooth surface and the probe is directed toward the soft tissue wall and attempted to proceed in an apical direction to bypass the obstruction.
4. When the obstruction is dental calculus, the periodontal probe will be moved more apically towards the bottom of the pocket.
5. When there is no obstruction, the advancement of the periodontal probe will be impeded by another "obstruction" that feels relatively soft, elastic and resilient: it is the junctional epithelium (Pattison et al., 1985). (Figure 3, Figure 4).

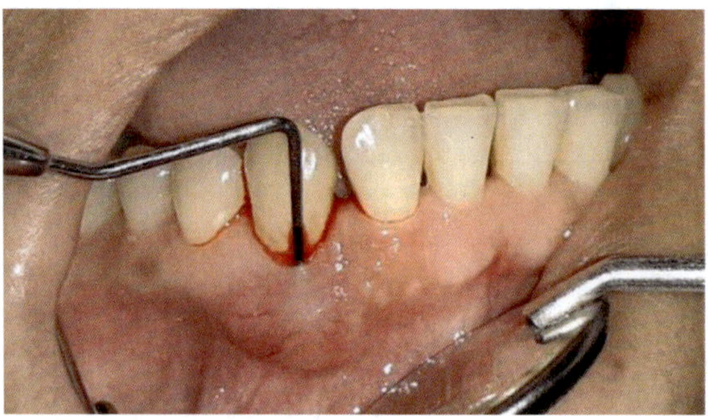

Figure 3. The periodontal probe enters 9 mm into the buccal face of the tooth 4.3. Moderate gingival bleeding is noted.

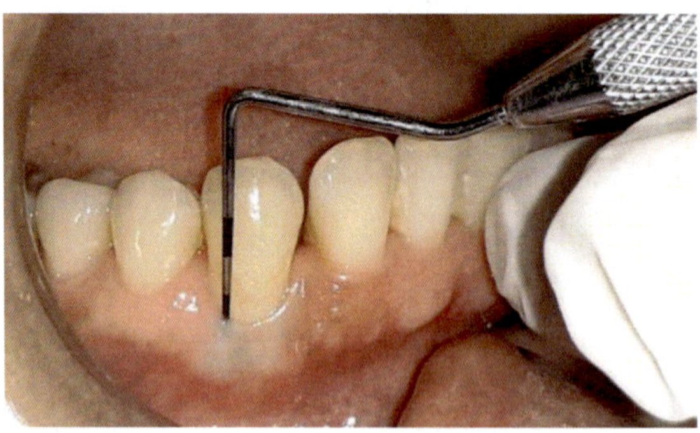

Figure 4. After periodontal treatment by root planing, the periodontal probe only enters 3mm.

Non-surgical Periodontal Treatment

Faced with periodontal disease, the primary approach to professional work is debridement of bacterial biofilm and dental calculus using non-surgical techniques and procedures.

Control of the bacterial biofilm lays the foundation for long-term success of the treatment. The lasting positive effects of periodontal therapy are within easy reach. As long as therapeutic compliance is achieved by the patient and we carry out effective control of the biofilm (Perry et al., 2012).

Scaling

The American Academy of Periodontics defines scaling as "an instrumentation of the crown and root surfaces of the teeth to remove plaque, dental calculus, and pigmentation from these surfaces." That is, it is the removal of the subgingival calculus (Academy of Periodontology, 2021).

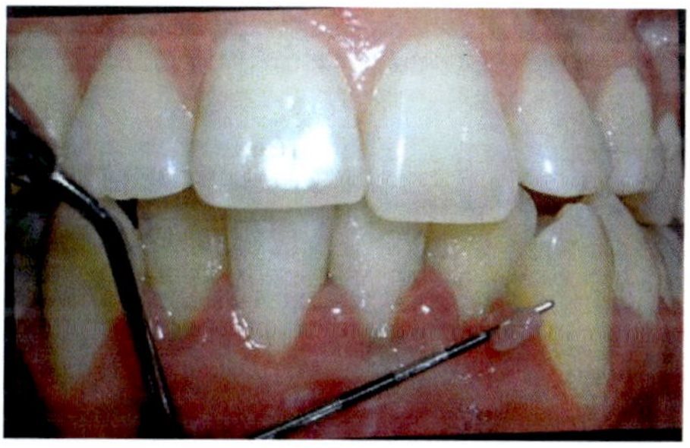

Figure 5. The gingival tissues are enlarged, have a reddish color and one papilla shows gingival growth, which becomes noticeable with the periodontal probe.

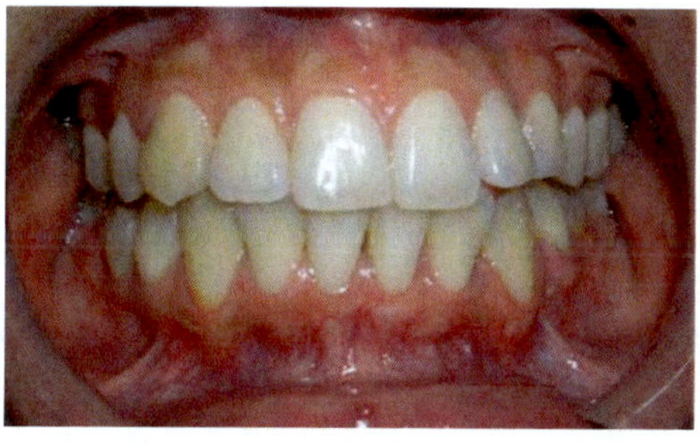

Figure 6. Gingival tissues are completely healthy after scaling and root planing treatment.

In the therapy process, after the scaling procedures, the success of the treatment is evaluated by exploring the uniformity of the roots in order to ensure the total removal of the dental calculus. It is carried out with sharp hand instruments or by sonic or ultrasonic instrumentation (Figure 5, Figure 6).

Root Planing

According to the American Academy of Periodontics, it is "the therapeutic procedure designed to remove cement or rough surface dentin, impregnated with dental calculus and/or contaminated with toxins or microorganisms" (Academy of Periodontology, 2021).

This procedure does not address identifiable dental calculus deposits but assumes the entire root surface associated with the periodontal pocket. The goal is to remove the superficial layer of cementum or dentin that may be impregnated with bacterial lipopolysaccharides (endotoxins) or dental calculus to create a hard, glassy surface (Matsuda et al., 2011).

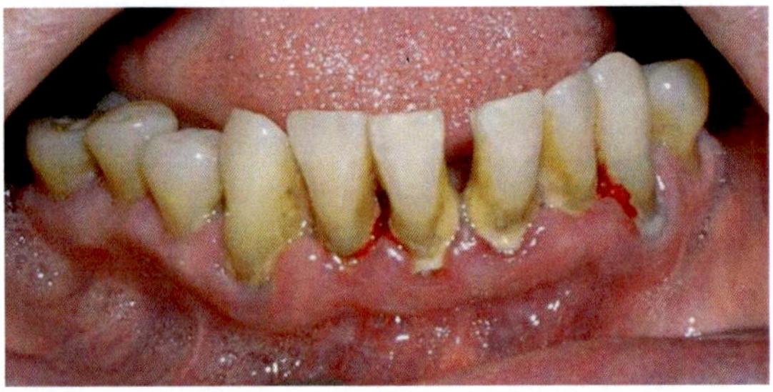

Figure 7. There is an abundant amount of supragingival calculus on the teeth of the lower anterior area and the gingival tissues àre inflamed.

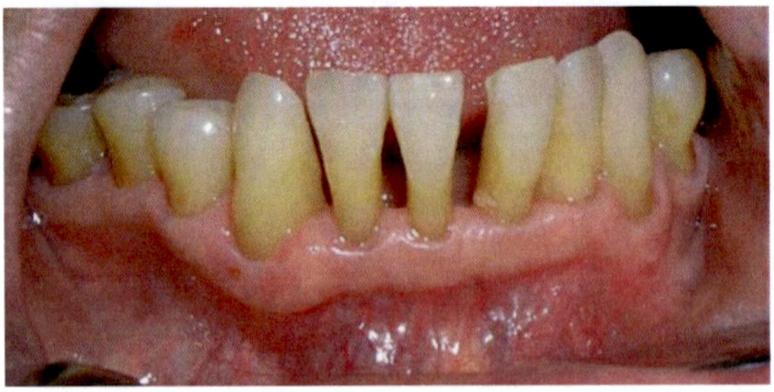

Figure 8. After periodontal treatment by scaling and root planing, the periodontal tissues present a healthy pink color.

When the professional perceives that the root surfaces are smooth and hard, he can feel confident that the treated pockets are free of deposits and contaminants embedded or adhered to the root surfaces (Loomer et al., 2004).

The difference between scaling and root planing is a matter of degrees. Root planing involves a concrete effort to instrument all areas of the root surfaces, and not just the identifiable dental calculus deposits. The terms non-surgical periodontal treatment, periodontal debridement, and the traditional terms scaling and root planing have been used together (Figure 7, Figure 8).

Periodontal Prophylaxis

It is a preventive procedure to eliminate local gingival irritants. The dental calculus is completely removed and root planing is continued. It is also called as oral prophylaxis, dental prophylaxis or just prophylaxis. Its goal is to help the patient maintain and preserve their periodontal health. A polishing procedure is performed and is known as prophylaxis, although this term does not correspond to it.

Goals of Non-surgical Periodontal Treatment

At the time of the appointment with the patient and the non-surgical periodontal treatment has been decided, it is essential to establish its immediate objectives and long-term objectives.

The immediate goals of treatment are to promote control of the bacterial biofilm. For this we must instrument all dental surfaces until they are completely clean and smooth. At the same time we must evaluate the work with the periodontal probe (Cobb, 1996).

Long-term treatment goals allow restoring periodontal health. Multiple appointments are needed to reach this goal. It is necessary to consider, within the process, sufficient time to monitor the recovery of periodontal tissues. Keep in mind that in the case of epithelium and connective tissue it takes several months (Cobb, 1996).

Scaling and root planing are the fundamental pillars of the non-surgical and non-pharmacological treatment of chronic periodontal diseases. Both therapeutic procedures decrease gingival inflammation and probing depth. This means a gain in clinical insertion in the majority of patients with periodontal pathologies (Cobb, 1996).

As the biofilm ages and calcifies, the organic matrix and bacteria also calcify. An interesting work by Kepic et al. demonstrated the presence of dental calculus residues in most of the treated teeth, between 45 and 60 minutes after the quadrant treatment time. Even when each piece was instrumented for 40 minutes, stone remains were regularly found in the deepest pockets. Completely clean surfaces were only obtained in pockets of 3 to 4 mm (Stambaugh et al., 1981; Nyman et al., 1988).

This indicates the thoroughness and thoroughness to responsibly perform scaling and root planing.

It also remains to be answered whether the root surfaces have to have a glassy appearance. Variation in texture appears to be acceptable as long as the dental calculus, which makes the surface rough and irregular, has been removed. Studies indicate that endotoxins do not penetrate deeply into cementum surfaces and that retained toxins are associated with inadvertent dental calculus and plaque, rather than diseased cementum (Beemsterboer et al., 2014).

The toxins are located on the root surfaces and are easily removed. Therefore, extensive root instrumentation is not required, beyond dental calculus and plaque removal. With this concept the fundamentals for smoothing roots in order to eliminate root roughness to obtain smooth and glassy surfaces are no longer valid.

Soft Tissue Healing after Scaling and Root Planing

Scaling and root planing cause some removal and disruption of the junctional epithelium and deeper connective tissue. When the junctional epithelium has been injured or separated from the root surfaces, as occurs during periodontal debridement, healing should be expected to take approximately a few weeks. The normal turnover of cells in the junctional epithelium, which migrate from the apical end to the coronal end, takes about 5 days. Healing of inflamed connective tissue is complex and requires the action of numerous cells and mediators. It takes considerably longer to complete than epithelial healing, up to several months (Badersten et al., 1984). (Figure 9, Figure 10).

Insertion of new connective tissue fibers is not a predictable outcome, but the development of an elongated attachment of the junctional epithelium may decrease probing measurements. Given the fragility of the connective tissues in the healing phase, after treatment periodontal probing should be avoided for about 4 weeks (Ciancio, 1989).

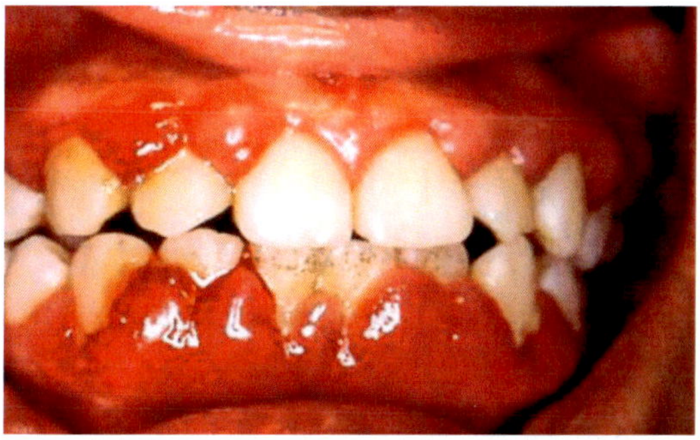

Figure 9. A 22-year-old female patient with multiple gingival enlargements in both jaws, in some areas the gingival growth is close to the incisal edge.

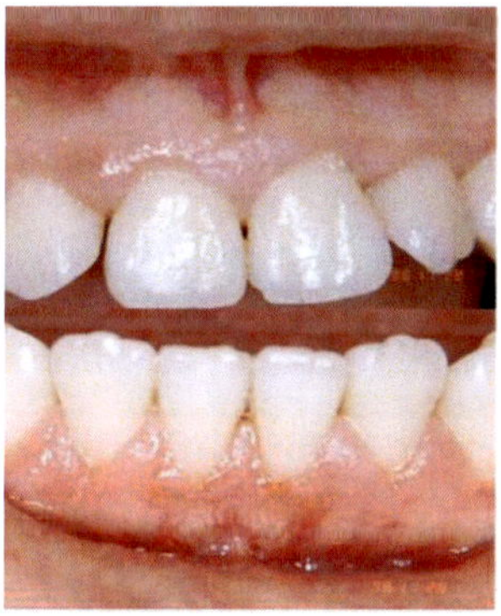

Figure 10. Periodontal treatment by scaling and root planing was performed and after 4 months her periodontal health was restored.

Scaling and root planing effectively reduce the number of biofilm bacteria from all treated areas. The number of microorganisms is dramatically reduced. The bacterial plaque varies from predominantly gram bacterial to gram positive, with fewer motile forms, especially spirochetes. This new microflora is very similar to healthy periodontal areas. The cycle can take up to 6 months to complete (Armitage, 1980).

Clinical Response

After thorough scaling and root planing, the expected clinical response is decreased periodontal pocket depth, attachment changes, recession, decreased bleeding, less tissue redness.

Minimal amounts of new bone are generated after non-surgical periodontal treatment. (Figure 11, Figure 12, Figure 13).

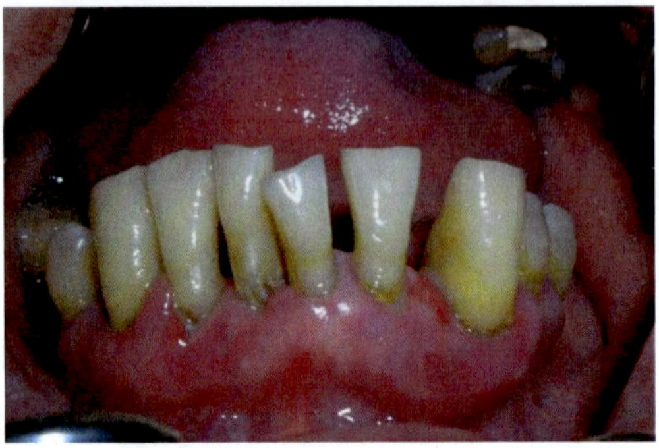

Figure 11. In the lower anterior area, supragingival calculus, reddened gingival tissue, very pronounced gaps and gingival retraction are observed.

The clinical results of several studies on non-surgical therapy reach these conclusions:

- The more superficial pockets decrease less than the deeper ones.
- Redness and bleeding dramatically decrease.
- Healing reaches a maximum response between 3 and 6 weeks after non-surgical periodontal treatment.

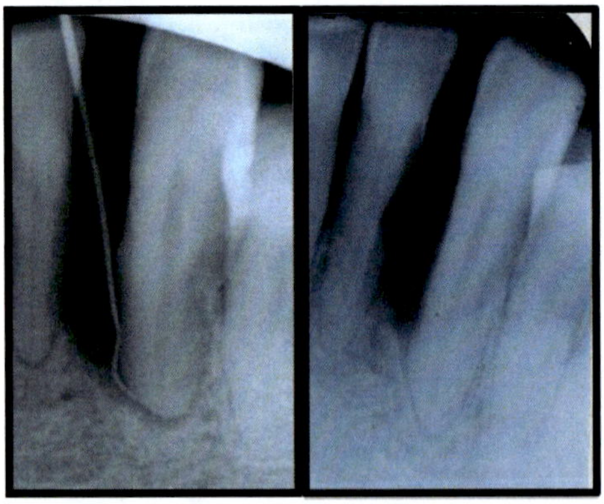

Figure 12. The same patient as in the previous figure. Image A shows bone destruction due to chronic periodontal disease that reaches the root apex. A gutta-percha cone makes it more evident. In image B after 5 months of treatment based on root planing, a new bone formation of many millimeters can be seen.

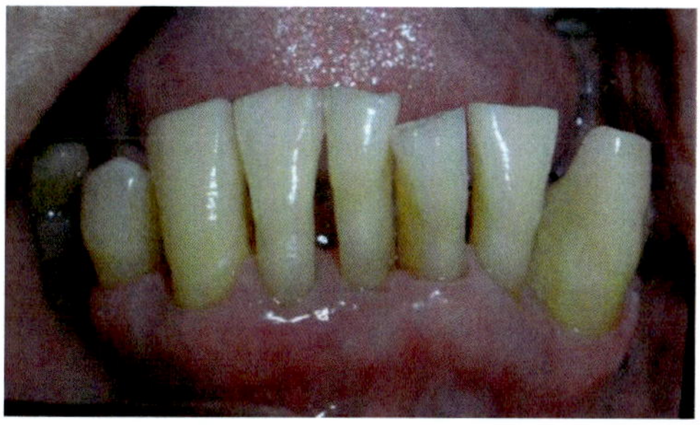

Figure 13. After scaling and root planing, recovery of the gingival tissues can be seen. In addition, a closure of the interproximal space is observed between pieces 3.1, 3.2 and 3.3.

Changes also appear after 12 months of treatment and at this time the clinical response stabilizes (Robertson et al., 1990).

The pockets with more than 7 mm depth are those that show a greater reduction, from 1.5 to 3 mm. In the studies analyzed, the number of bleeding areas decreased by 50% or more and gingival redness was reduced to practically zero, after non-surgical periodontal treatment (Badersten et al., 1984; Robertson et al., 1990).

Reassessment of probing depths several weeks after treatment probably reflects the histologic depth of pockets more accurately than probing depths prior to periodontal treatment. Reassessment of depths is important to monitor the long-term results of treatment and determine if more extensive periodontal treatment is needed. Results should not be reassessed earlier than 3 to 4 weeks after periodontal treatment (Ramfjord, 1990).

Studies have suggested that it may be advantageous to debride in fewer but longer sessions, as opposed to the traditional four-session approach to treating chronic periodontitis. In this way, the number of pathogens in the oral cavity would be reduced and bacterial repopulation in previously treated areas would be avoided (Mongerdini et al., 1999).

Unfortunately, this approach has not gained adherents. Some studies would add the administration of disinfectants to scaling and root planing to enhance the efficiency of biofilm removal and host response (Greenstein, 2002).

As with any periodontal treatment, other factors must be taken into account in establishing a treatment plan and lengthy treatment appointments may be indicated.

In a very interesting article "The enigma of scaling and root planing" Dr. Harpreet Singh Grover (2022) describes a reality about the academic training of scaling and root planing of undergraduate students. We quote some paragraphs.

> The dilemma of periodontics in 2022 stems from the lack of consensus on the practical training program in the undergraduate clinics of our dental students and the perceived lack of interest in it.
>
> Over the years, countless reasons have been advanced as to why training hours for scaling and root planing should not be cut short to be replaced by seemingly more interesting training methods such as ultrasonic scaling, minor surgical procedures, etc. Apparently, every self-respecting periodontist, whether he is a master or a pure clinician, must give his own reason for shooting down scaling and root planing. Many seem self-obsessed with the desire to be 'Messiah of change' slaying the demons of scaling and root planing, their minds in an archetypal Don Quixote haze.
>
> Instead of thinking about reducing hours, more time should be spent learning and mastering the fine art of manual scaling, developing a dexterity that will help to perform more effective sonic and ultrasonic scaling and a better level of mechanical debridement during periodontal surgeries as well.
>
> Let's not be naive. Declining interest is a stark reality, but it is equally stark that scapegoating (in this case scaling and root planing) is not the need of the hour. It goes without saying that undermining scaling and root planing means undermining dentistry itself (Grover, 2022).

This is probably a reality in many countries around the world. Academic changes should be made to improve periodontal care coverage for patients with periodontal disease. It would be a humanistic contribution for our peers and they can have a better quality of life.

51 years have passed since the publication of the following statement and yet this concept continues to be present: *All patients are not the same and not all plaque is the same, so there is an infinite variety of possible clinical responses. Richard Parker* (Charles et al., 1972).

Acknowledgments

To Professor Mg. Lily Cardich for the careful revision of the style of the original manuscript. Likewise, to Dra. Nuria L. Salas Garcia for her excellent work typing the original document.

Conclusion

The treatment of periodontal diseases in their different stages: initial, moderate and even in an advanced state, the recovery of periodontal health is achieved, using the periodontal treatment procedures of scaling and root planing. Dentists in general should receive better academic training on scaling and root planing in their pre-pregation studies, which will reduce high levels of periodontal disease and the world population will have a better quality of life.

References

Academy of Periodontology. 2021. *Glossary of Periodontal Terms*. USA: Chicago. https://members.perio.org/libraries/glossary?ssopc=1.

Adriaens PA, Adriaens LM. 2004. Effects of nonsurgical periodontal therapy on hard and soft tissues. *Periodontol* 2000. 36: 121-145. Accessed April 26, 2023. doi: 10.1111/j.1600-0757.2004.03676.x.

American Academy of Periodontology. 2011. Comprehensive periodontal therapy: a statement. *J Periodontol.* 2011; 82: 943-949. Accessed April 27, 2023. doi: https://doi.org/10.1902/jop.2011.117001o.

Armitage GC. 1980 *Biologic Basis of Periodontal Maintenance Therapy*. 33-115. Berkeley: Praxis.

Badersten A, Nilvéus R, Egelberg J. 1981. Effect of non-surgical periodontal therapy. I. Moderately advanced periodontitis. *J Clin Periodontol.* 8:57-72. Accessed April 26, 2023. doi: 10.1111/j.1600-051x.1981.tb02024.x.

Badersten A, Nilveus R, Egelberg J. 1984. Effect of non-surgival periodontal therapy. II. Serevely advanced periodontitis. *J Clin Periodontol.* 11:63-76. Accessed April 26, 2023. doi: 10.1111/j.1600-051x.1984.tb01309.x.

Badersten A, Nilveus R, Egelberg J. 1987. Four-year observations of basic periodontal therapy. *J Clin Periodontol.* 14: 438-444. Accessed April 26, 2023. doi: 10.1111/j.1600-051x.1987.tb02248.x.

Barletta L. 2009. En el tratamiento periodontal, porqué siempre el raspaje [In periodontal treatment, why always scaling?]. *Rev. Fundac.* Juan Jose Camaro; 14 (29): 26-28. Accessed April 27, 2023. https://pesquisa.bvsalud. org/portal/resource/pt/lil-528993.

Beemsterboer P, Perry D. 2014. Tratamiento periodontal no quirúrgico [Non-surgical periodontal treatment]. *Periodontología para el higienista dental.* 186-196. España: Elsevier.

Botero JE, Kucherbecker C, Duque A, Jaramillo A, Contretas A. 2015. Periodontal disease in children and adolescents of Latin America. *Periodontol* 2000. 67: 34-37. Accessed April 26, 2023. doi: 10.1111/prd.12072.

Brayer, WK, Mellonig, JT, Dunlap, RM, Marinak, KW, & Carson, RE. 1989. Scaling and root planning effectiveness: the effect of root surface access and operator experience. *J Periodontal.* 60:67-72. Accessed April 27, 2023. doi: 10.1902/jop.1989.60.1.67.

Carranza FA Jr, Soghe R, Newman MG, Valentin PL. 1983. Scanning and transmission electron microscopic study of tissue-invading microorganisms in localized juvenile periodontitis. *J Periodontol*. 54: 598-617. Accessed April 27, 2023. doi: 10.1902/jop. 1983.54.10.598.

Charles A. Amenta, Jr. Robert C. Brackett, Richard Parker. 1972. *Dialogue on preventive dentistry.* Chicago; March Publishing Company Ill.

Ciancio S. 1989. Non-surgical periodontal treatment. *Proceeding of the world workshop in clinical periodontics*. Chicago: American Academy of Periodontology.

Cobb CM. 1996. Non-surgical pocket therapy: mechanical. *Ann Periodontol*. 1(1):443-490. Consultado el 26 de abril de 2023. doi: 10.1902/annals.

Contreras A, Slots J. 1998. Active cytomegalovirus infection in human periodontitis. *Oral Microbiol Immunol.* 13: 225-230. Accessed April 26, 2023. doi: 10.1111/j.1399-302x.1998.tb00700.x.

Contreras RA. 2016. La promoción de la salud general y la salud oral: una estrategia conjunta [The promotion of general health and oral health: a joint strategy.]. *Revista Clínica de Periodoncia, Implantología y Rehabilitación Oral*. 9: 193-202. Consultado el 26 de abril de 2023. doi:.

Dentino A, Lee S, Mailhot J, Hefti, AF 2013. *Principles of periodontology, Periodontol* 2000. 61(1):16-53. Accessed April 26, 2023. doi: 10.1111/j.1600-0757.2011.00397.x.

Donlan, Costerton JW. Biofilms: Survival Mechanism of Clinically Relevant Microorganisms. *Clinical Microbiology Reviews* 2022, 167-93. Accessed April 26, 2023. https://pesquisa.bvsalud.org/portal/resource/pt/lil-794517.

Falotico PG, Farias RF. 2006. El surco gingival. Aspectos clínico y anatomofisiomicrio-biológicos [The gingival sulcus. Clinical and anatomophysiomicryobiological aspects]. *Odous científica*. 2: 16-26. Accessed April 27, 2023. http://servicio.bc.uc.edu.ve/odontologia/revista/v7n2/7-2-3.pdf.

Farias RF. 2003. Enfermedad periodontal y microorganismos periodontopatógenos [Periodontal disease and periodontopathogenic microorganisms]. *Revista Odous Cienfífica* 2003. 6: 12-30. Accessed April 27, 2023. http://scielo.sld.cu/scielo.php?script=sci_arttext&pid=S1029-30192012000700014.

Federación Europea de periodoncia. *El impacto socioeconómico de la periodontitis* [The socioeconomic impact of periodontitis]. 2021. https://www.efp.org/fileadmin/uploads/efp/Documents/Other_publications/TheEconomist_Infohraphic_Spain.pdf.

Greenstein G. 2002. Full-mouth therapy verses individual quadrant root planning: a critical commentary. *J Periodontol*. 73: 797-812. Accessed April 27, 2023. doi: 10.1902/jop. 2002.73.7.797.

Grover HS. 2022. The comundrum of scaling and root planning. *Indian Soc Periodontal.* 26 (3): 203. Accessed April 27, 2023. doi: 10.4103/jisp.jisp_128_22.

Hamada S, Amano A, Kimura S, Nakagawa I, Kawabata S, Morisaki I. 1998. The importance of fimbriae in the virulence and ecology of some oral bacteria. *Oral Microbiol Immunol.* 13: 129-138. Accessed April 26, 2023. doi: 0.1111/j.1399-302x. 1998.tb00724.x.

Jhonson RB, Streckfus KF, Daix, Fucci Ma. 1995. Protein recovery from several paper types used to collect gingival crevicular fluid. *J of Periodontol* Res. 34: 283-289. Accessed April 27, 2023. doi: 10.1111/j.1600-0765.1999.tb02255.x.

Liébana VJ. 2002 *Microbiología Oral. Microbiología periodontal.* 549-551. España: Ed Mc Graw Hill.

Listgarten MA. 1976. Structure of the microbial flora associated with periodontal health and disease in man. A light and electron microscopic study. *J Periodontol.* 47: 1-18. Accessed April 26, 2023. doi: 10.1902/jop.1976.47.1.1.

Loomer PM, Armitage GC. 2004. *Microbiology of periodontal diseases.* 69-84. USA: Elsevier Mosby.

Maita Véliz L, Ramos PD, Castañeda MMM, Galvez C., Maita Castañeda L, Navarro CC. 2016. Presencia de bacterias negro-pigmentadas en la pared de la bolsa periodontal al inicio y término de la terapia periodontal [Presence of black-pigmented bacteria in the wall of the periodontal pocket at the beginning and end of periodontal therapy]. *Odontología Sanmarquina.* 19(1): 47-51. Accessed April 26, 2023. doi: 10.15381/os.v19i1.12183.

Maita Véliz LV, Maita Castañeda LM. 2003. La enfermedad periodontal como factor de riesgo para nacimiento pretérmino [Periodontal disease as a risk factor for preterm birth]. *Odontología Sanmarquina.* 6 (12): 37-40. Accessed April 26, 2023. doi: https://doi.org/10.15381/os.v6i12.3954.

Maita Véliz, L. 1992. Diagnóstico precoz de la enfermedad periodontal en los niños y adolescentes mediante el índice de sangrado papilar de Mühlemann (PBI) [Early diagnosis of periodontal disease in children and adolescents using the Mühlemann Papillary Bleeding Index (PBI)]. *Avances en periodoncia.* 4: 47-54. Accessed April 27, 2023. https://dialnet.unirioja.es/servlet/articulo?codigo=5674950.

Maita-Véliz LV, Maita-Castañeda LM. 2022. Non-surgical treatment of periodontal diseases: Responsibilities of the dentist and the patient in periodontal therapy. Periodontology-New insights. *IntechOpen.* Accessed April 26, 2023. doi: 10.5772/intechopen.107332.

Marsh PD. 1994. Microbial ecology of dental plaque and its significance in health and disease. *Adv Dent Res.* 8: 263-271. Accessed April 26, 2023. doi: 10.1177/08959374940080022001.

Matsuda SA, Wilkins EM. 2011. *Nonsurgical periodontal instrumentation.* 608-638. Philadelphia: Lippincott Williams & Wilkins.

Mongardini, C, Steenberghe, DV, Dekeyser, C, & Quirynen, M. 1999. One stage full-verses partial-mouth disinfection in the treatment of chronic adult or generalized early-onset periodontitis I: Long-term clinical observation. *J Periodontol.* 70:632-645. Accessed April 27, 2023. doi: 10.1902/jop.1999.70.6.632.

Moore WEC, Ranney RR, Holdeman LV. 1982. *Subgingival microflora in periodontal disease: culture studies.* 13-26 Washington DC: American Society fon Microbiology.

Newman and Carranza's. 2018. *Clinical Periodontology.* 1985: Elsevier.

Newman MG, Saglie FR. 1984. The role of microorganisms in periodontal disease. *Glickman's Clinical periodontology.* 361 390. Philadelphia: W.B Saunders Co.

Nyman, S, Westfelt, E, Sarhed, G, & Karring, T. 1988. Role of diseased root cementum in healing following treatment of periodontal disease: a Clinical study. *J Clin Periodontol.* 15: 464-468. Accessed April 27, 2023. doi: 10.1111/j.1600-051x.1988.tb01601.x.

Oda S, Mitta H, Setoguchi T, Izumi Y, Ishikawa I. 2004. Current concepts and advances in manual and power-driven instrumentation. *Periodontol* 2000. 36: 45-58. Consultado el 26 de abril de 2023. doi: 10.1111/j.1600-0757.2004.03674.x.

Page RC, Kornman KS. 1997. The pathogenesis of human periodontitis: an introduction. *Periodontol* 2000. 14:9-11. Accessed April 26, 2023. doi: 10.1111/j.1600-0757.1997.tb00189.x.

Page RC, Offenbacher S, Schroeder HE, Seymour GJ, Kornman KS. 1997. Advances in the pathogenesis of periodontitis: Summary of developments, clinical implications and future directions. *Periodontol* 2000. 4:216-248. Accessed April 26, 2023. doi: 10.1111/j.1600-0757.1997.tb00199.x.

Pattison G, Pattison AM. 1985. *Instrumentación Periodontal. Un manual clínico* [*Periodontal Instrumentation. A clinical manual*]. España: Elsevier.

Perry DA, Beemsterboer PL, Esset G. 2014. *Periodontología para el higienista dental* [*Periodontology for the dental hygienist*]. 168-172. España: Elsevier.

Perry DA, Takei HH, et al. 2012. *Phase I, periodontal therapy*. 448-451. MO: Elsevier Sauders.

Petersilka GJ, Ehmke B, Flemmig T. 2002. Antimicrobial effects of mechanical debridement. *Periodontol* 2000. 28: 56-71. Accessed April 27, 2023. doi: 10.1034/j.1600-0757.2002.280103.x.

Ramfjord SP. 1990. Long-term assessment of periodontal surgery versus curettage or scaling and root planning. *Int J Technol Assess Health Care*. 6: 392-402. Accessed April 27, 2023. doi: 10.1017/s0266462300001008.

Robertson PB, Buchanan SA. 1990. *Wound healing after periodontal therapy*. 382-393. Louis, MO: Mosby.

Stambaugh RV, Dragoo M, Smith DM, Carasali, L. 1981. The limits of subgingival scaling. *Int J Periodont Res Dent*. 1: 30-41. Accessed April 27, 2023. https://pubmed.ncbi.nlm.nih.gov/7047434/.

Ting, M, Contreras, A, & Slots, J. 2000. Herpesvirus in localized juvenile periodontitis. *J Periodontal Res*. Accessed April 26, 2023. doi: 10.1034/j.1600-0765.2000.035001017.x.

Chapter 2

The Impact of Metal Pollution on Biodiversity

Samir Ghannem[1,*], PhD
Tahani El Ayari[1], PhD
Raja Ben Ahmed[2], MA
and Samir Touaylia[1], MC
[1]University of Carthage, Faculty of Sciences of Bizerte, Laboratory of Environment Bio-Monitoring, Bizerte, Tunisia
[2]University of El Manar, Faculty of Sciences of Tunis, Department of Biology, Ecology, Biology and Physiology of Aquatic Organisms Laboratory, Tunis, Tunisia

Abstract

Among the various contaminants, heavy metals are of particular concern due to their environmental persistence, biogeochemical recycling and ecological risks. They are considering among the most hazardous environmental pollutants, because they do not disintegrate with physical processes and therefore remain for longtime period. Through a process of biomagnification and bioaccumulation, heavy metals had a significant impact in biological systems. Thus, metal pollution can cause a decline in certain pollinating populations, difficulties for certain species to reproduce or to feed themselves, modify the physiology of organism, the anatomy and the characteristic of the biotope of local populations. The consequences of disturbances are detected at different levels, among others at the individual level, at the level of populations and at the level of the dynamics of biological communities. Populations, subject to pollution, seek to maintain their balance and limit disturbance, this by reducing competing species and therefore the number of taxa present.

[*] Corresponding Author's Email: ghan_samir@yahoo.fr.

In: Advances in Biology. Volume 4
Editor: Charles D. Grant
ISBN: 979-8-88697-999-2

Metal pollutants are widespread in air, soil and water causing a decline in fauna and flora worldwide. In this brief chapter, we have tried to develop a comprehensive understanding of the influence of metal pollution on ecosystem disturbance. We give examples of studies on the effects of pollutants on biodiversity.

Keywords: heavy metals, pollution, ecological risk, biodiversity decline

Introduction

Biodiversity, or biological diversity, is defined as the variety of life on earth in all its forms. It includes the number of species, their genetic variation and the interaction of these life forms within complex ecosystems (Ghannem, 2017). This biodiversity has been declining at an alarming rate in recent years. According to van Klink et al. (2020), terrestrial insect abundance shows an alarming rate of decline of ca. 1% every year for a century. Human activities, such as land use changes, climate change and pollution are the main cause. Heavy metal pollution is one of the main causes of environmental degradation. Metal pollution and the loss of biodiversity are two major ecological crises of this century, which threaten the functioning of ecosystems. The loss of biodiversity is a particularly serious threat to the functioning of ecosystems, which depend on the presence of multiple species to maintain themselves. According to Mahanta et al. (2022), the use of fertilizers and chemicals is at the origin of the decline of biodiversity in agricultural ecosystems since it eliminates a high number of insects acting as biological control species. However, the growth of metal pollution in the environment causes a negative impact on organisms and influences the arrangement and functioning of the soil ecosystem (Zhou and Song, 2004). Heavy metal pollution of soil is a major environmental problem in the last few decades owing to the rapid increase in urbanization and industrialization (Aydi et al., 2022; Ghannem et al., 2018a; Purushotham et al., 2012); however, in soil pollution, the major problem in the environment and metals exists either as separate entities or in combination in soil (Chandrasekaran et al., 2015). Metals are considering among the most hazardous environmental pollutants, because they do not disintegrate with physical processes and therefore remain for longtime period (Kasassi et al., 2008). Through a process of biomagnification and bioaccumulation, heavy metals had a significant impact in biological systems. It can affect the environment at different levels of the trophic chain and are potentially noxious to the ecosystem and humans (Manahan, 2000). It could damage different parts of plant as well as their biochemical and physiological functions of their components including cell membranes, mitochondrial, nuclei and DNA (Wang et al., 2012). Heavy metal pollution affects the quality of the atmosphere and water bodies, and threatens the health and life of microorganisms, terrestrial and aquatic flora and fauna (Monchanin et al., 2021). By contact, ingestion or inhalation toxic elements have a negative effect on biota and human health. Furthermore, due to the bioaccumulation and biomagnification of toxic metals, toxicity increases from low to high trophic levels in aquatic and terrestrial ecosystems will be most severely affected (Talukder et al., 2021; Yunus et al., 2020). Even metals that are necessary for the functioning of living organisms such as zinc, copper and chromium, their overconsumption can lead to a series of adverse effects (Krishnan et al., 2022).

Bioindication Concept

Bioindication is used in many fields, such as monitoring the quality of air, water, soil, monitoring biodiversity, monitoring the impact of human activities on ecosystems. However, bioindication allows a more global and more ecological approach to the assessment of the state of the environment, because it takes into account the interactions between different living organisms and their environment. This helps to better understand the cumulative effects of pollution and environmental changes on ecosystems. In addition, the ecological integrity of a natural environment is determined by its degree of chemical, physical and biological integrity. The degradation of one or more of these components generally occurs in the biocenosis, i.e., in different biological communities (MDDEP, 2011). In this sense, the morphological, behavioral, tissue or physiological alterations, thus the abundance, the reproductive success and the mortality of animal and plant species constitute observable and miserable variables reflecting the state of their habitat (Harrage, 2012, Kerckhove, 2012). Bioindication can also be used to assess the effects of pollution on ecosystems. For example, studying the distribution and migration of certain species can help predict the impact of pollution on biodiversity. Bioindication is based on the process of analyzing several biological indicators which is part of the biomonitoring of the quality of ecosystems (Chabot, 2014). However, bioindicators make it possible to characterize the present state of a key habitat, as well as the evolution of the state of the ecosystem over time (Bentata-Keddar, 2015). Consequently, they can thus constitute a means of controlling the effectiveness of the conservation measures put in place, and guide the necessary corrective measures, if necessary. They can also identify a source of disruption to the ecosystem (Kerckhove, 2012). The definition used by several authors such as Banaru and Perez (2010), A biological indicator (or bio-indicator) is an organism or a set of organisms which - by reference to biochemical, cytological, physiological, ethological or ecological variables - allows, in a practical and safe way, to characterize the state of an ecosystem or an eco-complex and to highlight as early as possible their modifications.

Biodiversity Bioindicators

Bio-indicators can also be used as biodiversity indicators, also called community change bio-indicators (Kerckhove, 2012). Since indicator taxa are sensitive to changes in the ecosystem, changes that could have impacts on local biodiversity, they can also be used as surrogates to assess the risks of impact on this biodiversity (Kerckhove, 2012). Biodiversity bio-indicators are organisms used to assess the biological diversity of a given environment. They make it possible to measure the richness and variability of species present in an ecosystem, as well as their spatial distribution and abundance. Biodiversity bio-indicators can be animal, plant or microbial species. The species chosen as bio-indicators are often those which are endemic to a given environment or which have an important function in the ecosystem. For example, the Cicindelinae serve as indicators of biodiversity (Dajoz, 2002), similarly, bees are often used as bio-indicators of grassland biodiversity, as they play a key role in plant pollination. Biodiversity bio-indicators can be used to assess the impact of human activities on the environment, such as pollution, land use change or habitat fragmentation. They can also be used to assess the effectiveness of biodiversity conservation measures, such as ecological restoration, creation of protected areas or ecosystem management. Biodiversity bio-indicators have several advantages over traditional biodiversity assessment methods, as they are often more sensitive to environmental change and easier to measure. However, their use requires a thorough knowledge of the ecology of the species used and a global approach to the measurement of biodiversity.

Pollution-Sensitive and Pollutant-Resistant Species

Hundreds of species can survive exclusively in very little polluted environments. These are then qualified as polluosensitive. On the other hand, pollutant-resistant species can survive in environments exposed to higher levels of pollution (MEDDE, 2012). Pollutant resistant species, on the other hand, are organisms that can tolerate high levels of environmental pollutants. Their presence in a given environment can indicate pollution, but this does not necessarily mean that the pollution is very important. For example, some plant species can grow in soil contaminated with heavy metals, which may indicate pollution, but it does not mean that the soil is hazardous to human health or the environment. It is important to note that the use of pollutant-sensitive and pollutant-resistant species in bioindication should be used with care and in combination with other measurement methods. Indeed, pollutant-sensitive species can be eliminated from an environment before the pollution is detected, while pollutant-resistant species can adapt to high levels of pollution and continue to grow even if the environment is hazardous to health human or for the environment.

Biotic Index

The biotic index corresponds to a quantitative assessment of the quality of an ecosystem based on the relative abundance of species, the presence/absence of indicator species, or according to the extent of changes observed in bioindicators (Melquiote, 2008). The biotic index is an important tool for the management and conservation of ecosystems. It makes it possible to measure the effects of human activities on the biodiversity of the ecosystem and to take measures to reduce the negative impacts on species. The biotic index can also be used to assess the biodiversity of primarily aquatic ecosystems. Indeed, the biodiversity of aquatic ecosystems is often linked to water quality, as pollution and other disturbances can affect the diversity and abundance of the species that live there. In general, the more different species present in the ecosystem, the higher the biodiversity is considered to be. However, it is important to note that the biotic index does not take into account the genetic diversity of the species present, which is also an important aspect of biodiversity. Moreover, it does not measure interactions between species and ecosystem functions, which are also important components of biodiversity. Thus, although the biotic index can give a useful indication of the biodiversity of aquatic ecosystems, it should not be used as an exclusive indicator of biodiversity. Other measures, such as genetic and functional diversity, must also be taken into account to obtain a more complete assessment of the biodiversity of an ecosystem.

Ecological Indices

Relative Abundance

It corresponds to the ratio of the number of a given species to the sum of the numbers of all the species occupying the same station. The mean relative abundance (Pi) is evaluated as follows:

Pi = n/Ns with:

n = number of individuals of a species (i) collected in a given station;

Ns = total number of individuals of all species collected in the same station.

Species Richness

$d = (S-1) / \log_2 N$

S: number of species at each station;

N: total number of species.

The Shannon-Wiener Index (H´)

"H´" is expressed in Bits and defined by the following expression:

$$H' = - \sum (Pi \times Log_2 Pi)$$

"H´" varies between 0 (only one species is abundant) and Log_2 S (all species have the same abundance). This index can assess the spatial and/or temporal diversity of a given stand.

Equitability (Es)

It provides information on the degree of biotopic diversity and the degree of equitable distribution of species between the surveyed sites. Its value varies from 0 (dominance of a single species) to 1 (even distribution of individuals in the species):

$$Es = H' / Log_2 S$$

Effect of Metals

Invertebrate Biodiversity

Heavy metals can have negative effects on the biodiversity of invertebrates by altering their development, behavior, reproduction and survival. Referring to Tovar-Sánchez et al. (2018), heavy metal pollution can threaten biodiversity. However, exposure to heavy metals exerts alterations across levels of biological organization. In addition, population distribution and abundance, community structure, and ecosystem dynamics may be altered. Exposure to heavy metals can affect invertebrates in different ways. Aquatic invertebrates, for example, can be exposed to heavy metals through contaminated water, while terrestrial invertebrates can be exposed through their food or habitat. Heavy metals can affect their growth and development, which can lead to morphological abnormalities, malformations and reduced body size. These changes therefore represent a threat to biodiversity. In addition, metal pollution can also affect the reproduction of invertebrates by receiving the fecundity and quality of eggs and larvae, which can have effects on the survival of populations. They can also disrupt the behavior of invertebrates, which can affect their ability to reproduce and avoid predators and consequently a decline in the composition and diversity of invertebrate communities in ecosystems. Some invertebrates are more sensitive to heavy metals than others, which can lead to decreased species diversity and altered community structure. For example, the presence of heavy metals in water can reduce aquatic invertebrate species diversity and lead to an increase in heavy metal-tolerant species. Hackstein (1988), reports that Cd metal in water causes increased mortality in the population of *Gammarus tigrinus* species. Spurgeon and Hopkin (1996) found a negative correlation between the distance to the source of pollution and the contents of the elements (Pb, Cd, Zn and Cu). Similarly, a negative correlation was noticed between the values of heavy metals and the absolute abundance of six species of earthworms (Phyllum Annelida: *L. rubellus, L. cataneus, L. terrestris, A. rosea, A. caliginosa, A. chlorotica*). Essentially, the species *A. rosea*, *A. caliginosa* and *A. chlorotica* were absent from the most polluted sites (Pb, Cd, Zn and Cu) (Spurgeon and Hopkin, 1999). They observed that the relative abundance and the Shannon-Weiner diversity index were lower at sites closest to the pollution source and higher at sites furthest from the pollution source. However, the dominance index (Berger-Parker) was higher in sites close to the pollution source because *L. rubellus*, *L. castaneus* and *L. terrestris* were the dominant species in these sites. According to Nahmani and Lavelle (2002), the absolute and relative abundances of organisms from the Phyla Annelida and Arthopoda community were lower in the most polluted sites. However, the density of *A. caliginosa* earthworms was higher in unpolluted sites and a total disappearance in heavily polluted sites. According to Creamer et al. (2008), in an effort to investigate the response of soil invertebrate diversity and activity to elevated soil metal values resulting from sewage sludge application, observed no disturbance of the soil abundance or overall diversity (based on the Shannon-Weiner index) of the invertebrate community harvested for metal values within limits. For values above the limits, sensitivity was noticed at the level of the specific richness, while the overall diversity was not affected. Earthworm abundance was significantly reduced at total Cu values equal to and greater than 176 mg kg^{-1}, while nematode and enchytraeid abundance was sensitive to high Cu and Zn values. Referring to Acevedo-Whitehouse and Duffus (2009) pollution intensities could have negative effects on the restriction of the distribution area or even the disappearance of certain species. In a polluted environment morphological, behavioral, tissue or physiological alterations, as well as the abundance, reproductive success and mortality of animal species are observable and measurable variables reflecting the state of their habitat (Bednarska and Świątek, 2016; Kaiser et al., 2001).

Insect Biodiversity

Heavy metals can have harmful effects on insects, which play an important role in terrestrial ecosystems as predators, pollinators and decomposers (Ghannem et al., 2016). The effects of heavy metals on insects depend on factors such as the type of metal, the dose, the duration of exposure, the stage of development of the insect and the route of exposure (by contact, ingestion or inhalation) (Ghannem et al., 2018a). Heavy metals can affect the development, behavior, reproduction and survival of insects. For example, lead can reduce insect fecundity and ability to feed, while cadmium can affect insect growth and reproduction. Mercury can also affect insect survival and ability to move (Ghannem et al., 2018b). This type of pollution therefore leads to a decline in biodiversity. Insect biodiversity is essential for maintaining the balance of terrestrial ecosystems, as insects play an important role in pollination, the breakdown of organic matter and the regulation of populations of other animal and plant species. Heavy metals can affect insect biodiversity in different ways. For example, insects can be exposed to heavy metals through their food or prey, which can affect their health and reproduction. Insects that feed on contaminated plants can accumulate heavy metals in their bodies, which can affect their behavior and survival (Ghannem et al., 2018a).

Heavy metals can also affect insect reproduction by affecting fecundity in females and sperm quality in males. In addition, metal pollution can affect insect diversity by invading the population of certain species or changing the composition of the insect community in a given ecosystem. For example, heavy metals can reduce the population of pollinating insects, which can affect fruit and vegetable production in crops. Heavy metals can also affect insects that feed on organic matter by altering their ability to decompose organic matter, which can have consequences for the quality of soil and the biodiversity of its microorganisms. According to Pragya et al. (2015), the humoral immunity of invertebrates could be affected by the stress of toxic elements, and the response of the humoral immunity of insects was found to affect the susceptibility of insects to entomopathogenic microorganisms (Afify et al., 2019; Pragya et al., 2015). In addition, insects reduce their feeding or decrease their digestion and absorption as a result of heavy metals accumulated in food (Li et al., 2018). Referring to Long et al. (2015), a reduction in intermediate carbohydrate metabolism, was observed in chironomid (*Chironomus tepperi*), including glucose 6-phosphate, fructose 6-phosphate and disaccharides following exposure to Zn stress. Thus, to better understand the effect of heavy metal stress on insect disease resistance and their corresponding regulatory mechanisms, studies on several insect species and different types of toxic elements should be done (Jiang et al., 2021). Co-exposure of honey bees (*A. mellifera*) to cadmium and copper resulted in increased developmental time, high mortality, and decreased food intake and response to sucrose (Di et al., 2020). Referring to Osman et al. (2015), the study of the impact of Cadmium on species of Tenebrionidae reveals a decrease in population density, a reduction in body weight, an increase in the mortality rate and an increase in the sex ratio insects. Moreover, the results also revealed a striking decrease in the body length of polluted insects with a marked increase in the percentage of deformed gonads and digestive tract. Therefore, heavy metal pollution is one of the factors that threaten biodiversity.

Bird Biodiversity

Metals can have negative effects on the biodiversity of birds. Birds can be exposed to toxic metals through their diet, the water they drink, the air they breathe, or their environment. Metals can affect birds in a number of ways, including disrupting their feelings, weakening their immune systems, affecting their behavior and reproduction, altering their physiology, and causing disease. Heavy metals such as lead, mercury and cadmium are of particular concern as they have toxic effects at low doses. Tan et al. (2009) and Chastel (2014) indicate that even moderate concentrations of mercury or POPs are associated with disruption of the functioning of certain hormones (corticosterone, LH, prolactin). However, exposure to these contaminants significantly affects the selective value of individuals (suppression of reproduction, alteration of parental behavior, long-term decline in fertility) and alters the very dynamics of populations. The translation of environmental cues (climate, resources) into ecological responses (laying date) occurs via a cascade of hormonal processes, which probably determine the degree of adaptation to environmental changes. In this context, exposure to certain pollutants such as POPs and mercury could lead to greater sensitivity to stress factors (modification of habitats, reduction in food resources, disturbance) and alter the ability of individuals to cope with them. In addition, exposure to metals can also have indirect effects on the biodiversity of birds by affecting their habitat or their prey. For example, metal contamination of soils can lead to a decrease in the availability of insects, which are a food source for many birds. The effects of metals on the biodiversity of birds can vary depending on the amount and duration of exposure, the type of metal, the age and health of the bird, and its habitat and diet. The birds most sensitive to metals are often those that are in danger of extinction or whose populations are already in decline. It is therefore important to monitor the levels of metals in the environment and to take measures to reduce the exposure of birds to toxic metals in order to preserve bird biodiversity.

Microbial Activity

The soil is a compartment of the ecosystem with multiple functions such as: support for living organisms and microorganisms, reservoir of organic and mineral matter, place of transformations, purifying system, regulator of flows and exchanges in biogeochemical cycles (cycle of nitrogen, carbon, etc. (Gobat et al., 1998). Many organizations are involved in carrying out these functions. Different types of microorganisms are represented in the soil where they play a crucial role in the biogeochemical cycles of elements. Heavy metals are natural elements that can be toxic to soil microorganisms at high concentrations. The effect of heavy metals on soil microorganisms depends on several factors, such as metal type, concentration, pH, and soil texture. Heavy metals such as lead, cadmium and mercury can affect the growth, reproduction and diversity of soil microorganisms, which can lead to reduced soil fertility. Soil microorganisms can be affected in different ways by heavy metals. Heavy metals can affect the permeability of cell membranes, DNA structure, enzyme production and cellular respiration. Heavy metals can also affect the interaction between soil microorganisms and plants by altering the availability of nutrients to plants. However, it is important to note that soil microorganisms may also have the ability to tolerate and adapt to certain heavy metals, which may lead to changes in the soil microbial community. In addition, some species of soil microorganisms are able to degrade certain heavy metals, which may contribute to their removal from the soil. The impacts of heavy metals on microbial communities can be approached in various ways: density (colony forming units, CFU), size, community structure (genetic and functional) and also enzymatic activity. Several studies show that the bacterial biomass of a soil tends to decrease following contamination by a metal (Ekelund et al., 2003; Kelly et al., 1999; Konopka et al., 1999). Moreover, Giller et al. (1998) estimate that, even in the long term and for low heavy metal contents, microorganisms are not able to maintain a biomass equivalent to that of unpolluted soil. Studies of the impact of heavy metals on the diversity of the bacterial community in the soil have shown mainly a negative influence (Hinojosa et al., 2005; Moffett et al., 2003; Sandaa et al., 2001). Bacteria and fungi isolated from polluted soils are more tolerant to heavy heavy metal contamination than those from unpolluted soils (Mertens et al., 2006; Doelman et al., 1994;). The addition of heavy metals therefore causes the disappearance of the most sensitive populations and subsequently the adaptation of the most resistant populations. Thus, balances can shift and dominance can be reversed (Díaz-Roviña and Bååth, 1996). It has been shown that changes in communities following the presence of copper could lead to an increase in the relative proportion of Gram-negative bacteria (Proteobacteria) (Turpeinen et al., 2004) or, on the contrary, Gram-positive bacteria (*Firmicutes*) (Ekelund et al. al., 2003).

Discussion

Our chapter of the literature on heavy metal shows many negative effects of these metal pollutants on biodiversity. The harmful effects of the high-risk element cause serious consequences, sometimes difficult to assess, such as impaired fertility (springtail: (Crouau and Pinelli, 2008); earthworm : (Konečný et al., 2014); developmental defects (fly: (Nascarella et al., 2003); moth: (van Ooik et al., 2007); ant: (Skaldina et al., 2018), resistance to pathogens (ant: (Sorvari et al., 2007); honey bee: (Polykretis et al., 2016)) and also altered feeding behavior (aphid: (Stolpe et al., 2017); honey bee: (Burden et al., 2019)). All these effects lead to a decrease in biodiversity. Referring to Osman et al. (2015), the study of the impact of cadmium on species of Tenebrionidae, reveals a decrease in population density, a reduction in body weight, an increase in the mortality rate and an increase in the sex ratio insects. Moreover, the results also revealed a striking decrease in the body length of polluted insects with a marked increase in the percentage of deformed gonads and digestive tract. According to Drobne (1997), toxic elements, particularly zinc, can disturb molting cycles, alter the digestive glands in isopods and can affect food consumption, reproduction and the composition of the community. Previous studies indicate that metal pollution has severe consequences on pollinator behavior by affecting foraging activity, food perception and memory abilities necessary for energetic foraging (Monchanin et al., 2021; Burden et al., 2019; Xun et al., 2018; Sivakoff, 2017. The results obtained by Kheirallah and El-Samad (2019) on tenebrionid beetles (*Blaps polycresta* and *Trachyderma hispida*) inhabiting polluted soil shows several abnormalities in oocytes and trophocytes. These baseline abnormalities included exfoliation and vacuolation of follicular epithelium, vacuolated trophocyst, nuclear abnormalities, and morphological changes in cytoplasmic organelles. Co-exposure of honey bees (*A. mellifera*) to cadmium and copper resulted in increased developmental time, high mortality, and decreased food intake and response to sucrose (Di et al., 2020). It is clear the severe effect of metal toxicity on species and consequently the negative impact on biodiversity.

Conclusion

Metal pollution is one of the major risks in today's world. This is a topical problem that concerns all regions concerned with maintaining their biodiversity at a high level of quality. However, metals can have negative effects on biodiversity in general by disrupting ecosystems and the populations of animal and plant species. Heavy metals such as lead, mercury and cadmium are of particular concern as they can accumulate in the environment and have toxic effects at low doses. Exposure to metals can also have indirect effects on biodiversity by affecting habitats and food chains. To preserve biodiversity, it is therefore crucial to monitor the levels of metals in the environment and to take measures to reduce the exposure of organisms to toxic metals.

Disclaimer

None

References

Acevedo-Whitehouse K, Duffus AL. Effects of environmental change on wildlife health. *Philosophical Transactions of the Royal Society B: Biological Sciences* (2009) 364(1534):3429-3438.

Afify A, Ghezawy NA, Ali MM, El-Tayeb TA. Effects of copper chlorophylline and magnesium chlorophylline on the humeral immune response, ultrastructural changes of midgut and DNA damage to Culex pipiens larvae. *African Entomology* (2019) 27(1):146-158.

Aydi A, Ghannem S, Nasri A, Hessine R, Mezni A. Evaluation of heavy metals contamination and pollution indices levels in surface sediments of the Bizerte coastal line, Tunisia. *Marine Pollution Bulletin* (2022) 184:114171.

Banaru D, Perez T. *Bioindicateurs, Biomarqueurs, Notes de cours* [*Bioindicators, Biomarkers, Course notes*]. Université de Marseille, Paris; 2010.

Bednarska AJ, Świątek Z. Subcellular partitioning of cadmium and zinc in mealworm beetle (*Tenebrio molitor*) larvae exposed to metal-contaminated flour. *Ecotoxicology and Environmental Safety* (2016) 133:82-89.

Bentata-Keddar I. *Evaluation de la contamination métallique par trois métaux traces (Cd, Ni et Zn) du rouget de vase Mullus barbatus (L., 1758) pêché au niveau de la côte occidentale algérienne* [*Evaluation of metallic contamination by three trace metals (Cd, Ni and Zn) of red mullet Mullus barbatus (L., 1758) fished on the Algerian west coast*]. Doctoral dissertation, Thesis of Magister, University of Oran, Algeria; 2015

Burden CM, Morgan MO, Hladun KR, Amdam GV, Trumble JJ, Smith BH. Acute sublethal exposure to toxic heavy metals alters honeybee (*Apis mellifera*) feeding behavior. *Scientific reports* (2019) 9(1):1-10.

Chabot B. *Les facteurs de sélection des bio-indicateurs de la qualité des écosystèmes aquatiques: Elaboration d'un outil d'aide à la décision*. Mémoire de maîtrise en environnement [*Factors for selecting bioindicators of the quality of aquatic ecosystems: Development of a decision support tool*. Master's thesis in environment]. Université de Sherbrooke, Québec; 2014.

Chandrasekaran A, Ravisankar R, Harikrishnan N, Satapathy KK, Prasad MVR, Kanagasabapathy KV. Multivariate statistical analysis of heavy metal concentration in soils of Yelagiri Hills, Tamilnadu, India–Spectroscopical approach. *Spectrochimica Acta Part A: Molecular and Biomolecular Spectroscopy* (2015) 137:589-600.

Chastel O. *Stress, hormones et effort parental chez les oiseaux: une approche évolutive* [*Stress, hormones and parental effort in birds: an evolutionary approach*]. Doctoral dissertation, Université de La Rochelle; 2014.

Creamer RE, Rimmer DL, Black HIJ. Do elevated soil concentrations of metals affect the diversity and activity of soil invertebrates in the long-term? *Soil Use and Management* (2008) 24(1):37-46.

Crouau Y, Pinelli E. Comparative ecotoxicity of three polluted industrial soils for the Collembola *Folsomia candida. Ecotoxicology and Environmental safety* (2008) 71(3):643-649.

Dajoz R. *Les Coléoptères Carabidés et Ténébrionidés : écologie et biologie.* [*Carabid and Tenebrionid beetles: ecology and biology.*] Lavoisier, Paris; 2002.

Di N, Zhang K, Hladun KR, Rust M, Chen YF, Zhu ZY, Liu, T-X, Trumble JT. Joint effects of cadmium and copper on *Apis mellifera* forgers and larvae. *Comparative Biochemistry and Physiology Part C: Toxicology &* Pharmacology (2020) 237: 108839.

Díaz-Roviña M, Bååth E. Development of metal tolerance in soil bacterial communities exposed to experimentally increased metal levels. *Appl. Environ. Microbiol* (1996) 62:2970-2977.

Doelman P, Jansen E, Michels M, Van Til M. Effects of heavy metals in soil on microbial diversity and activity as shown by the sensitivity-resistance index, an ecologically relevant parameter. *Biol. Fertil. Soils* (1994) 17:177-184.

Drobne D. Terrestrial isopods—a good choice for toxicity testing of pollutants in the terrestrial environment. *Environmental Toxicology and Chemistry: An International Journal* (1997) 16(6):1159-1164.

Ekelund F, Olsson S, Johansen A. Changes in the succession and diversity of protozoan and microbial populations in soil spiked with a range of copper concentrations. *Soil Biology and Biochemistry* (2003) *35*(11):1507-1516.

Ghannem S, Beldi A, Amar WB, Boumaiza M. Contribution à La Connaissance des Coleopteres terrestres du parc national de l'Ichkeul [Contribution to Knowledge of the Terrestrial Coleoptera of the Ichkeul National Park]. *Revue F.S.B XV* (2017).

Ghannem S, Khazri A, Sellami B, Boumaiza M. Assessment of heavy metal contamination in soil and *Chlaenius (Chlaeniellus) olivieri* (Coleoptera, Carabidae) in the vicinity of a textile factory near Ras Jbel (Bizerte, Tunisia). *Environmental Earth Sciences* (2016) 75:1-10.

Ghannem S, Touaylia S, Boumaiza M. Beetles (Insecta: Coleoptera) as bioindicators of the assessment of environmental pollution. *Human and Ecological Risk Assessment: An International Journal* (2018b) 24(2):456-464.

Ghannem S, Touaylia S, Bejaoui M. Assessment of trace metals contamination in soil, leaf litter and leaf beetles (Coleoptera, Chrysomelidae) in the vicinity of a metallurgical factory near Menzel Bourguiba (Tunisia). *Human and Ecological Risk Assessment: An International Journal* (2018a) 24(4):991-1002.

Giller KE, Witter E, Mcgrath SP. Toxicity of heavy metals to microorganisms and microbial processes in agricultural soils: a review. *Soil biology and biochemistry* (1998) 30(10-11):1389-1414.

Gobat JM, Aragno M, Matthey W. Le sol vivant: Bases de pédologie [The living soil: Basics of pedology]. *Biologie des Sols* (1998) 415-515.

Hackstein E. Changes in the Population Dynamics of *Gammarus tigrinus* Sexton (Crustacea: Amphipoda) as Expression of Sublethal Effects by Reciprocal Interactions of Temperature and Cadmium Enriched Food. *Internationale Revue der Gesamten Hydrobiologie IGHYAZ* (1988) 73(2).

Harrag. *Bio-indication: Concepts et Biosurveillance, Notes de cours* [*Bioindication: Concepts and Biomonitoring, Course Notes*]. Sétif, Université, 7p. htpp.//fr.Calameo.com/cpm/001058687a447e5d0152 (page consultée le 8 novembre 2013); 2012.

Hinojosa MB, Carreira JA, García-Ruíz R, Dick RP. Microbial Response to Heavy Metal–Polluted Soils: Community Analysis from Phospholipid-Linked Fatty Acids and Ester-Linked Fatty Acids Extracts. *Journal of Environmental Quality* (2005) 34(5): 1789-1800.

Jiang D, Tan M, Guo Q, Yan S. Transfer of heavy metal along food chain: a mini-review on insect susceptibility to entomopathogenic microorganisms under heavy metal stress. *Pest Management Science* (2021) 77(3):1115-1120.

Kaiser J. *Bioindicators and biomarkers of environmental pollution and risk assessment.* Hartford, Science Publishers; 2001.

Kasassi A, Rakimbei P, Karagiannidis A, Zabaniotou A, Tsiouvaras K, Nastis A, Tzafeiropoulou K. Soil contamination by heavy metals: Measurements from a closed unlined landfill. *Bioresource Technology* (2008) 99(18):8578-8584.

Kelly JJ, Häggblom M, Tate Iii RL. Changes in soil microbial communities over time resulting from one time application of zinc: a laboratory microcosm study. *Soil Biology and Biochemistry* (1999) 31(10): 1455-1465.

Kerckhove O. *Espèces ou association d'espèces de poissons en tant que bioindicateur de l'état de santé des récifs* coralliens (Mémoire de maîtrise). [*Fish species or association of fish species as a bioindicator of the health status of coral reefs* (Master's thesis)]. Sherbrooke University, Sherbrooke, Québec; 2012.

Kheirallah DA, El-Samad LM. Oogenesis anomalies induced by heavy metal contamination in two tenebrionid beetles (*Blaps polycresta* and *Trachyderma hispida*). *Folia Biologica (Kraków)* (2019) 67(1):9-23.

Konečný L, Ettler V, Kristiansen SM, Amorim MJB, Kříbek B, Mihaljevič M, Scott-Fordsmand JJ. Response of *Enchytraeus crypticus* worms to high metal levels in tropical soils polluted by copper smelting. *Journal of Geochemical Exploration* (2014) 144:427-432.

Konopka A, Zakharova T, Bischoff M, Oliver L, Nakatsu C, Turco RF. Microbial biomass and activity in lead-contaminated soil. *Applied and Environmental Microbiology* (1999) *65*(5) 2256-2259.

Krishnan K, Saion EB, Yap CK, Chong MY, Nadia AS. Détermination des éléments traces dans les échantillons de sédiments en utilisant l'analyse par activation neutronique [Determination of trace elements in sediment samples using neutron activation analysis]. *J. Exp. Biol. Agric. Sei* (2022) 10 (1):21-31.

Long SM, Tull DL, Jeppe KJ, De Souza DP, Dayalan S, Pettigrove VJ, McConville, MJ, Hoffmann AA. A multi-platform metabolomics approach demonstrates changes in energy metabolism and the transsulfuration pathway in *Chironomus tepperi* following exposure to zinc. *Aquatic Toxicology* (2015) 162:54-65.

Mahanta DK, Singh MK, Sujatha GS, Teja KSS. Insects as Ecological and Biological Indicators. *Vigyan Varta* (2022) 3(5): 136-143.

Manahan SE. *Environmenal chemistry,* Lewis, Boca Raton; 2000.

MDDEP. *Protocole d'échantillonnage des macroinvertébrés benthiques d'eau douce du Québec, cours d'eau peu profonds à substrat meuble, direction du suivi de l'état de l'environnement* [*Sampling protocol for freshwater benthic macroinvertebrates in Quebec, shallow watercourses with soft substrate, direction for monitoring the state of the environment*], MDDEP; 2011.

Melquiot P. Norme AFNOR pour la biosurveillance de l'environnement [AFNOR standard for environmental biomonitoring]. In A*ctualités - news environnement, Pollution de l'air- Air ambiant.* [*News - environment news, Air pollution- Ambient air*]. http://www.actualités- news environnement.com/19257- norme-afnor-biosurveillance -environnement.html; 2008)

Moffett BF, Nicholson FA, Uwakwe NC, Chambers BJ, Harris JA, Hill TCJ. Zinc contamination decreases the bacterial diversity of agricultural soil. *FEMS Microbiol. Ecol* (2003) 43:13-19.

Monchanin C, Devaud JM, Barron AB, Lihoreau M. Current permissible levels of metal pollutants harm terrestrial invertebrates. *Science of the Total Environment* (2021) 779-146398.

Nahmani J, Lavelle P. Effects of heavy metal pollution on soil macrofauna in a grassland of Northern France. *European Journal of Soil Biology* (2002) 38(3-4): 297-300.

Nascarella MA, Stoffolano Jr, JG, Stanek III, EJ, Kostecki PT, Calabrese EJ. Hormesis and stage specific toxicity induced by cadmium in an insect model, the queen blowfly, *Phormia regina* Meig. *Environmental Pollution* (2003) 124(2):257-262.

Osman W, M El-Samad L, Mokhamer EH, El-Touhamy A, Shonouda M. Ecological, morphological, and histological studies on *Blaps polycresta* (Coleoptera: Tenebrionidae) as biomonitors of cadmium soil pollution. *Environmental Science and Pollution Research* (2015) 22(18):14104-14115.

Polykretis P, Delfino G, Petrocelli I, Cervo R, Tanteri G, Montori G, Perito, B, Branca, JJV, Morucci, G, Gulisano M. Evidence of immunocompetence reduction induced by cadmiumhoneybeese in honey bees (*Apis mellifera*). *Environmental Pollution (*2016) 218:826-834.

Pragya P, Shukla AK, Murthy RC, Abdin MZ, Kar Chowdhuri D. Characterization of the effect of Cr (VI) on humoral innate immunity using *Drosophila melanogaster*. *Environmental toxicology* (2015) 30(11):1285-1296.

Purushotham D, Lone MA, Rashid M, Rao AN, Ahmed S. Deciphering heavy metal contamination zones in soils of a granitic terrain of southern India using factor analysis and GIS. *Journal of earth system science* (2012) 121:1059-1070.

Sandaa RA, Torsvik, V, Enger Θ. Influence of long-term heavy metal contaminations on microbial communities in soil. *Soil Biol. Biochem* (2001) 33:287-295.

Sivakoff FS, Gardiner MM. Soil lead contamination decreases bee visit duration at sunflowers. *Urban Ecosystems* (2017) 20(6):1221-1228.

Skaldina O, Peräniemi S, Sorvari J. Ants and their nests as indicators for industrial heavy metal contamination. *Environmental Pollution* (2018) 240:574-581.

Sorvari J, Rantala LM, Rantala MJ, Hakkarainen H, Eeva T. Heavy metal pollution disturbs immune response in wild ante populations. *Environmental Pollution* (2007) 145(1):324-328.

Spurgeon DJ, Hopkin SP. Seasonal variation in the abundance, biomass and biodiversity of earthworms in soils contaminated with metal emissions from a primary smelting works. *Journal of Applied Ecology* (1999) 36(1):173-183.

Spurgeon DJ, Hopkin SP. The effects of metal contamination on earthworm populations around a smelting works: quantifying species effects. *Applied Soil Ecology* (1996) *4*(2):147-160.

Stolpe C, Krämer U, Müller C. Heavy metal (hyper) accumulation in leaves of *Arabidopsis halleri* is accompanied by a reduced performance of herbivores and shifts in leaf glucosinolate and element concentrations. *Environmental and experimental botany* (2017) 133:78-86.

Talukder R, Rabbi MH, Baharim NB, Carnicelli S. Source identification and ecological risk assessment of heavy metal pollution in sediments of Setiu wetland, Malaysia. *Environmental Forensics* (2022) 23(1-2):241-254.

Tan SW, Meiller JC, Mahaffey KR. Les effets endocriniens du mercure chez les humains et la faune [The endocrine effects of mercury in humans and wildlife]. *Revues critiques en toxicologie* (2009) 39(3):228-269.

Tovar-Sánchez E, Hernández-Plata I, Martinez MS, Valencia-Cuevas L, Galante PM. Heavy metal pollution as a biodiversity threat. *Heavy Metals* (2018) 383.

Turpeinen R, Kairesalo T, Häggblom MM. Microbial community structure and activity in arsenic, chromium and copper contaminated soils. *FEMS Microbiol. Ecol.* (2004) 47:39-50

Van Klink R, Bowler DE, Gongalsky KB, Swengel AB, Gentile A, Chase JM. Meta-analysis reveals declines in terrestrial but increases in freshwater insect abundances. *Science* (2020) 368(6489):417-420.

Van Ooik T, Rantala MJ, Saloniemi I. Diet-mediated effects of heavy metal pollution on growth and immune response in the geometrid moth *Epirrita autumnata*. *Environmental Pollution* (2007) 145(1):348-354.

Wang C, Liu S, Zhao Q, Deng L, Dong S. Spatial variation and contamination assessment of heavy metals in sediments in the Manwan Reservoir, Lancang River. *Ecotoxicology and Environmental Safety* (2012). 82:32-39.

Xun E, Zhang Y, Zhao J, Guo J. Heavy metals in nectar modify behaviors of pollinators and nectar robbers: Consequences for plant fitness. *Environmental Pollution* (2018) 242:1166-1175.

Yunus K, Zuraidah MA, John A. Un examen de l'accumulation de métaux lourds dans les sédiments côtiers de la Malaisie péninsulaire [A review of heavy metal accumulation in coastal sediments of Peninsular Malaysia]. *Écoféminisme et changement climatique* (2020) 1(1):21-35.

Zhou QX, Song YF. Remediation of contaminated soils: principles and methods. *Science, Beijing* (2004) 1-489.

Chapter 3

A Transcriptome Analysis of the Model Cyanobacterium *Synechococcus elongatus* PCC7942 under Severe Salinity and Antibiotic Stress Conditions

Hakuto Kageyama[1,2,*]
Takashi Hibino[1,2]
and Rungaroon Waditee-Sirisattha[3,†]
[1]Graduate School of Environmental and Human Sciences, Meijo University, Shiogamaguchi, Tempaku-ku, Nagoya, Aichi, Japan
[2]Department of Chemistry, Faculty of Science and Technology, Meijo University, Shiogamaguchi, Tempaku-ku, Nagoya, Aichi, Japan
[3]Department of Microbiology, Faculty of Science, Chulalongkorn University, Phayathai Road, Pathumwan, Bangkok, Thailand

Abstract

Intracellular responses such as regulatory changes in metabolic pathways under severe abiotic stress conditions in cyanobacteria are of interest. In this study, transcriptome analyses were conducted using the cyanobacterium *Synechococcus elongatus* PCC7942 exposed to severe salinity and antibiotic stresses. Salinity is one of the important abiotic stresses for cyanobacteria. Here, we treated the cyanobacterium with severe salt stress (0.4 M NaCl) and performed transcriptome analysis using RNA sequencing. As a result, the expression of 370 genes out of a total of 2661 genes was induced significantly for at least 3 days, and the

* Corresponding Author's Email: kageyama@meijo-u.ac.jp.
† Corresponding Author's Email: Rungaroon.W@chula.ac.th.

In: Advances in Biology. Volume 4
Editor: Charles D. Grant
ISBN: 979-8-88697-999-2

expression of 115 genes was markedly decreased. On the other hand, the protein translation system is essential molecular machinery. Therefore, antibiotics with protein synthesis inhibitory activity suppress the growth of microorganisms and cells. Here, we treated the cyanobacterium with streptomycin, which selectively binds to bacterial ribosomes, at a concentration of 50 μg/mL. Transcriptome analysis revealed that the expression of 206 genes and 569 genes were significantly increased and decreased, respectively. Genes whose expression was altered by each stress treatment showed considerable overlap. Specifically, 119 genes were overlapping in those with upregulation, and 99 genes were overlapping in those with downregulation. These results suggest that severe salt stress conditions might cause the inhibition of protein synthesis in this cyanobacterium. Although the relationship between the functions of the proteins encoded by the expression-changed genes and the stress resistance mechanisms is unknown, these data will be useful for the future investigation of the environmental stress resistance mechanism of cyanobacteria.

Keywords: antibiotic stress, cyanobacteria, salt stress, transcriptome analysis

Introduction

Cyanobacteria are oxygenic, photosynthetic microorganisms and major biomass producers on Earth (Häder et al., 2007). They can utilize atmospheric carbon dioxide as a major source of carbon. The central carbon metabolism in cyanobacteria includes the Calvin cycle, glycolysis, the pentose phosphate pathway, and the tricarboxylic acid cycle (Xiong et al., 2015). It has been shown that cyanobacteria participate in critical biogeochemical cycles and represent around 30% of the global primary production, especially in central oceans (Hagemann, 2011). In addition, due to their nitrogen-fixing ability, cyanobacteria are regarded to supply combined nitrogen to ecosystems, including both aquatic and terrestrial ecosystems (Häder et al., 2007). Cyanobacteria are ubiquitous on Earth, including harsh environments such as hot springs, volcanic islands, deserts, highlands, polar regions, and salt lakes (Mataloni & Komárek, 2004; Whitton & Potts, 2000). To survive the natural environment, cyanobacteria are thought to have acquired molecular mechanisms that enable them to sense and adapt to environmental stress conditions (Waditee-Sirisattha & Kageyama, 2022a).

We have been conducting research on the salt stress tolerance mechanism of cyanobacteria (Waditee-Sirisattha & Kageyama, 2023). So far, we have investigated the salt stress response mechanisms of the biosynthetic pathways of glycine betaine (GB) and mycosporine-2-glycine (M2G), which act as osmoprotectants, using the halotolerant cyanobacterium *Halothece* sp. PCC7418 (hereafter, *Halothece* 7418) (Waditee-Sirisattha et al., 2015; Waditee-Sirisattha et al., 2014; Waditee et al., 2005; Waditee et al., 2003). Osmoprotectant molecules are low-molecular-weight compounds synthesized to adapt cells to the osmotic pressure of a high-salt external environment. Both GB and M2G require glycine during their biosynthesis. Therefore, large amounts of glycine must be prepared for *Halothece* 7418 to adapt to high-salinity environments. Recently, by performing transcriptome analysis by RNA-seq, we found that five metabolic pathways, which include shikimate, pentose phosphate, amino acid biosynthetic, one-carbon metabolic, and central carbon metabolic pathways, cooperate to supply precursors of osmotically compatible solutes including glycine (Waditee-Sirisattha & Kageyama, 2022b). Besides the biosynthesis of osmoprotectants, we found using RNA-seq-based transcriptome analysis that salt stress induced the expression of genes involved in ion homeostasis and energy production (Waditee-Sirisattha et al., 2022). Proteins encoded by these genes are thought to contribute to adaptation to salt stress environments. Thus, omics approaches, such as transcriptomic and proteomic analyses, are useful for understanding molecular regulatory networks in cyanobacterial cells.

When analyzing the salt stress response by transcriptome analysis, it is common to apply a salt concentration treatment that allows the cells to live without dying. In the case of *Halothece* 7418, the growth range is approximately 0.2 to 4.0 M NaCl (Garcia-Pichel et al., 1998), so we usually culture *Halothece* 7418 at 0.5 M and increase the concentration to 2.0–2.5 M when salt stress is applied (Kageyama et al., 2011; Patipong et al., 2019a; Patipong et al., 2019b; Waditee-Sirisattha et al., 2022; Waditee-Sirisattha & Kageyama, 2022b; Waditee-Sirisattha et al., 2015; Waditee-Sirisattha et al., 2014). In the case of the salt-sensitive freshwater cyanobacterium *Synechococcus elongatus* PCC7942 (hereafter, *Synechococcus* 7942), there is a report that added approximately 0.3 M NaCl to the medium and analyzed the effects of salt stress on global gene expression using the Sanger sequencing method (Dong et al., 2023). In that report, when 0.3 M NaCl was applied for 4 days, the genes related to “nitrogen metabolism,” “energy metabolism,” “ABC transporters,” “membrane transport,” “O-antigen nucleotide sugar biosynthesis,” “metabolism,” “sulfur metabolism,” “peptidoglycan biosynthesis and degradation proteins,” “oxidative phosphorylation,” “transporters,” “photosynthesis proteins,” “amino sugar and nucleotide sugar metabolism," and "biosynthesis of other secondary metabolites" were reported to be upregulated in *Synechococcus* 7942 (Dong et al., 2023). Thus, salt stress could induce the expression of various functional genes in *Synechococcus* 7942. On the other hand, if salt is present in excess of viable concentrations, the cyanobacteria cannot grow and would die. Intracellular responses such as regulatory changes in metabolic pathways that occur during the process of cell death due to salt stress are of interest, but to our knowledge, no such analytical reports exist. Therefore, in this study, we performed a comprehensive analysis of gene expression under salt conditions in which the cells were not able to grow, using the salt-sensitive *Synechococcus* 7942. In *Synechococcus* 7942, which was subjected to salt stress treatment with a growth-stopping concentration of 0.4 M NaCl, 370 and 115 of the 2,661 genes examined were remarkably upregulated and downregulated, respectively.

In addition, we investigated the effect of antibiotic stress, using streptomycin, on global gene expression of *Synechococcus* 7942 to compare it with the effect of salt stress. To our knowledge, transcriptome analysis using *Synechococcus* 7942 treated with antibiotics such as streptomycin has not been reported. Streptomycin is an aminoglycoside antibiotic that inhibits the growth of prokaryotic cells by associating with the 30S ribosomal subunit (Sharma et al., 2007). Therefore, the addition of streptomycin to the medium at high concentrations would subject cyanobacteria to protein synthesis inhibitory stress. In addition to translation inhibition, antibiotics have other effects on cells. In Gram-negative bacteria, membrane permeability is believed to be involved in physiological adaptations to antibiotic stress. In *Escherichia coli*, it has been reported that the expression of genes involved in membrane permeability was altered under antibiotic stress conditions (Viveiros et al., 2007). Antibiotic stress also altered membrane permeability in the cyanobacterium *Fremyella diplosiphon* (Yalcin et al., 2022). In addition, there is a report that treatment with antibiotics increased the intracellular reactive oxygen species (ROS) (Yalcin et al., 2022). Excessive intracellular production of ROS can lead to oxidative damage that ultimately leads to cell death (Du et al., 2018). In *Synechococcus* 7942, whose growth was stopped by the addition of streptomycin at a concentration of 50 μg/ml, 206 and 569 of the 2,661 genes examined were remarkably upregulated and downregulated, respectively. In this chapter, we provide the results of these comprehensive transcriptome analyses.

Materials and Methods

Culture Conditions for Cyanobacteria

Synechococcus elongatus PCC7942 was grown photoautotrophically in BG11 medium under continuous light conditions by a light-emitting diode of light intensity equivalent to 70 μmol m^{-2} S^{-1} at 30°C. The growth of *Synechococcus* 7942 was monitored by measuring the absorbance at 730 nm using a UV/VIS photospectrometer UVMINI-1240 (Shimadzu, Kyoto, Japan).

Stress Treatments on Cyanobacteria

Before stress treatments, *Synechococcus* 7942 cells were grown photoautotropically in BG11 medium until exponential phase (OD_{730} ~ 0.45). Cells were then harvested by centrifugation at 3400 rpm at 25°C for 15 min. The collected cells were transferred into fresh medium containing streptomycin at a concentration of 50 μg/mL or NaCl at a concentration of 0.3, 0.4, or 0.5 M.

Extraction of Total RNA from Cyanobacteria

For RNA-seq analysis, *Synechococcus* 7942 cells were exposed to antibiotic stress with 50 μg/mL of streptomycin or salt stress with 0.4 M NaCl. The cells were collected 24 and 72 h after starting the stress treatments. Cells from before subjecting to the stress treatments were used as control. Biological duplicated samples were prepared and stored at –80°C until RNA extraction. Total RNA was extracted using TRIzol® LS reagent (Thermo Fisher Scientific, Inc., Waltham, MA, USA), according to the manufacturer's protocol. DNase I treatment was carried out to eliminate DNA contamination by using RNase-free recombinant DNase I (Takara Bio, Shiga, Japan). After treatment, DNase I was inactivated by the addition of ethylenediaminetetraacetic acid at a concentration of 24 mM and heat treatment at 80°C for 2 min. Then, RNA was recovered from the reaction mixture by ethanol precipitation. The RNA pellet was reconstituted with 50 μl of nuclease-free water (Takara Bio). RNA concentration was determined spectrophotometrically (BioSpec-nano, Kyoto, Japan) and visually confirmed using agarose gel electrophoresis.

RNA-Seq Analysis

RNA-seq was conducted by Bioengineering Lab. Co., Ltd., Kanagawa, Japan. The concentration of total RNA was measured using a Quantus Fluorometer and the QuantiFluor RNA System (Promega, Madison, WI, USA). The quality was then confirmed using the 5200 Fragment Analyzer System and the Agilent HS RNA Kit (Agilent Technologies, Inc., Santa Clara, CA, USA). Libraries were prepared using the MGIEasy RNA Directional Library Prep Set (MGI Tech Co., Ltd., Shenzhen, China), according to the manufacturer's protocol, following the removal of ribosomal RNA by using riboPOOL (siTOOLs Biotech, Munich, Germany). The concentration of RNA in each library was measured by utilizing a Qubit 3.0 Fluorometer and the dsDNA HS Assay Kit (Thermo Fisher Scientific, Inc.). Then, the quality of the libraries was confirmed using a Fragment Analyzer, the dsDNA 915 Reagent Kit/Agilent 2100 Bioanalyzer, and the High Sensitivity DNA Kit (Agilent Technologies). Following circularization using the MGIEasy Circularization Kit (MGI Tech Co., Ltd.), DNA nanoballs were prepared by utilizing the DNBSEQ-G400 RS High Throughput Sequencing Kit (MGI Tech Co., Ltd.). RNA sequencing was performed on a DNBSEQ-G400 (MGI Tech Co., Ltd.), using paired-end 200-bp sequencing.

Extraction of Up- and Down-regulated Genes

To identify differentially expressed genes (DEGs) under stress conditions, we used the average transcripts per million (TPM) values. In our previous analysis using *Halothece* 7418, log2(fold change (FC)) ≥ 2 and log2(FC) ≤ -1 were used as the threshold value for highly upregulated and downregulated genes, respectively. Two independent experiments were performed, and the genes that showed expression changes exceeding the threshold in both experiments were extracted as DEGs. Figure 1 shows scatter plots of transcriptional expression data for replicates of each condition demonstrating a high degree of correlation.

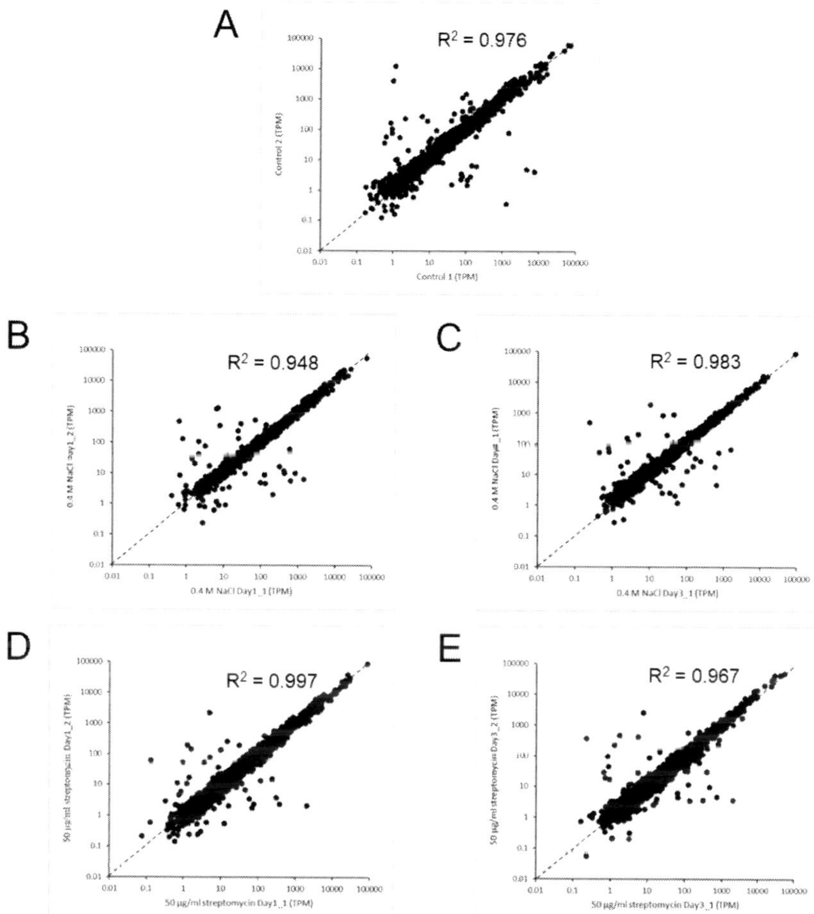

Figure 1. Scatter plots of transcriptional expression data for replicates of each condition. A) Control. B) Day 1 under 0.4 M NaCl stress. C) Day 3 under 0.4 M NaCl stress. D) Day 1 under antibiotic stress with 50 µg/ml of streptomycin. E) Day 3 under antibiotic stress.

Availability of RNA-Seq Data

Sequence data from the RNA-seq analysis has been deposited in the Gene Expression Omnibus (GEO), https://www.ncbi.nlm.nih.gov/geo/ (accession no. GSE225426).

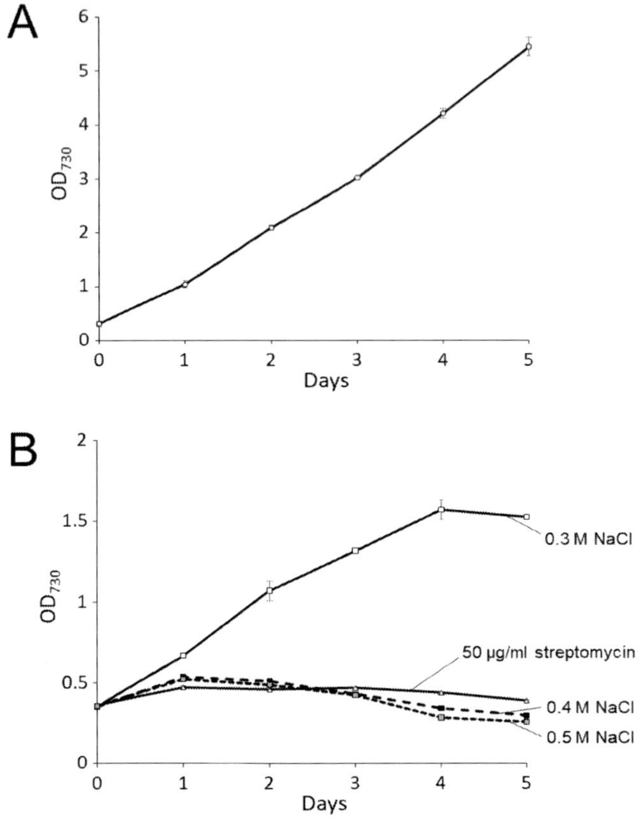

Figure 2. Effects of streptomycin and NaCl on the growth of *Synechococcus* 7942 cells. A) *Synechococcus* 7942 cells were cultivated in BG11 medium. B) *Synechococcus* 7942 cells were cultivated in BG11 containing NaCl or streptomycin. Each value shows the average of three independent measurements. Error bars show the standard errors (SE).

Results and Discussion

Effects of NaCl and Streptomycin on the Growth of *Synechococcus* 7942 Cells

First, we examined the relationship between growth and salt concentration to determine the strength of salt stress to apply to *Synechococcus* 7942 cells used for RNA-seq analysis. As shown in Figure 2A, *Synechococcus* 7942 cells continued to grow in normal BG11 medium for 5 days from the start of measurement, and the OD_{730} value exceeded 5 on day 5. On the other hand, Figure 2B shows that the growth of *Synechococcus* 7942 was suppressed in a concentration-dependent manner by salt stress. Under the 0.3 M NaCl condition, the OD_{730} value exceeded 1.5 on day 4, and no further increase was observed. Under the 0.4 M and 0.5 M NaCl conditions, a slight increase was observed on day 1, but the OD_{730} value gradually decreased thereafter. Thus, salinity with NaCl concentrations above 0.4 M seemed to stress the cells severely.

In the case of antibiotic stress using streptomycin, it was reported that the half maximal effective concentration (EC_{50}) of streptomycin for the cyanobacterium *Microcystis aeruginosa* was 0.28 μg/ml (Qian et al., 2012). Therefore, in this study, growth of *Synechococcus* 7942 was examined by adding streptomycin at a concentration of 50 μg/ml, which is considered to exhibit a sufficient protein synthesis inhibitory effect. As a result, under such antibiotic stress, time-course changes in OD_{730} values similar to those under 0.4 M and 0.5 M NaCl stress conditions were observed (Figure 2B).

Given the above observations, it was decided to subject *Synechococcus* 7942 to antibiotic stress and/or salt stress using 0.4 M NaCl and 50 μg/ml streptomycin to perform the RNA-seq analysis. These stress conditions were thought to strongly suppress cell proliferation.

Global Regulation of Gene Expressions in *Synechococcus* 7942 Cells under Stress Conditions

Effects of Salt Stress and Antibiotic Stress on Global Gene Expression in Synechococcus 7942

To explore the global gene expression profiles in *Synechococcus* 7942 cells exposed to fatal stress conditions, comparative transcriptomic analysis using RNA-seq was performed. We analyzed the total RNA samples prepared from stress-treated cells on days 1 and 3. In Figure 3, the number of DEGs from each sample is shown as a Venn diagram. Upregulated DEGs were more numerous on day 1 than on day 3 under salt stress conditions (Figure 3A). This suggests that many genes respond early to salt stress treatment. There were 370 DEGs whose expression was overlapping on both days. Antibiotic stress with streptomycin increased the expression of 260 and 288 genes on days 1 and 3, respectively (Figure 3A). Among the DEGs whose expression was elevated on both days, 206 genes were overlapping. On the other hand, regarding downregulated DEGs, the expression of 126 genes was suppressed on day 1 under salt stress conditions, and the number increased to 398 genes on day 3 (Figure 3B). It is thought that the global transcriptional activity was reduced in *Synechococcus* 7942 during the death process under the relatively strong salt stress environment of 0.4 M NaCl. Of the 126 genes detected on day 1, 115 genes were also detected on day 3. Under antibiotic stress conditions with streptomycin, suppression of many gene expressions was observed. On days 1 and 3, 678 and 702 genes were downregulated DEGs, respectively (Figure 3B).

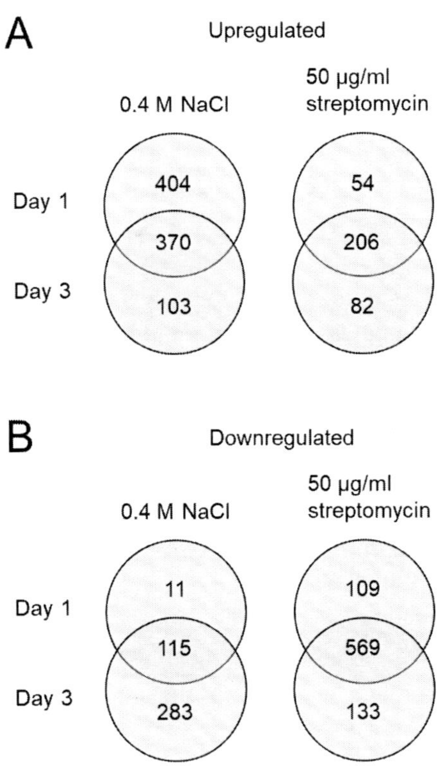

Figure 3. Venn diagram of the DEGs. A) Upregulated DEGs. The total number of upregulated DEGs was 774 on day 1 under salt stress, 473 on day 3 under salt stress, 260 on day 1 under antibiotic stress, and 288 on day 3 under antibiotic stress, respectively. B) Downregulated DEGs. The total number of downregulated DEGs was 126 on day 1 under salt stress, 398 on day 3 under salt stress, 678 on day 1 under antibiotic stress, and 702 on day 3 under antibiotic stress.

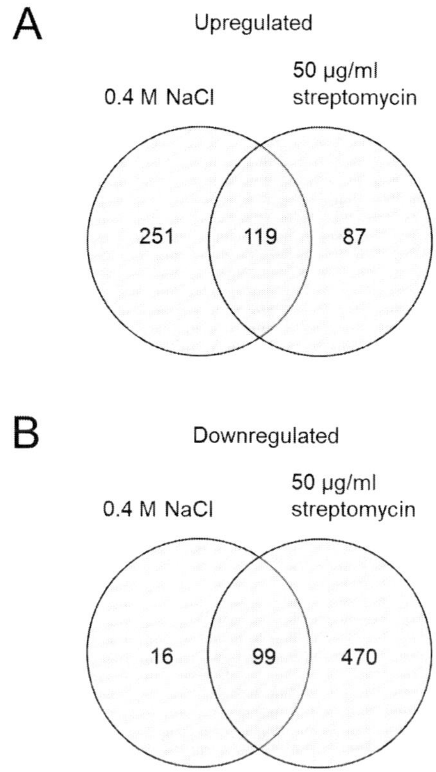

Figure 4. Venn diagram of the DEGs detected on both day 1 and day 3 under each stress condition. A) Upregulated DEGs. The total number of upregulated DEGs was 370 under salt stress, 206 under antibiotic stress, and 119 under both stresses. B) Downregulated DEGs. The total number of downregulated DEGs was 115 under salt stress, 569 under antibiotic stress, and 99 under both stresses.

Figure 4 shows a Venn diagram of the number of DEGs detected on both day 1 and day 3 under each stress condition. In the case of upregulated DEGs, 119 genes were common to salt stress and antibiotic stress (Figure 4A). Approximately 30–40% of the DEGs detected under each stress condition were overlapping, suggesting that common molecular mechanisms are involved in the response to these stress conditions. It is possible that 0.4 M NaCl stress reduced the activity of the protein translation system, resulting in a response similar to antibiotic stress. On the other hand, it has also been suggested that salt stress activates the translation system in cyanobacteria. For example, genes encoding ribosomal proteins are strongly induced by salt stress in the moderate halotolerant cyanobacterium *Synechocystis* sp. PCC6803

(Kanesaki et al., 2002). In the salt-tolerant cyanobacterium *Halothece* 7418, we recently observed the increased expression of genes encoding elongation factors as well as genes for ribosomal proteins (Waditee-Sirisattha et al., 2022). Furthermore, Billis et al. demonstrated that the *Synpcc2203-2205* operon comprising genes related to translation was upregulated in *Synechococcus* 7942 under the 0.2 M NaCl stress condition (Billis et al., 2014). However, these genes were not detected as highly upregulated DEGs in our experiment with 0.4 M NaCl. Given these observations, it was suggested that protein synthesis activity might be activated under moderate salt stress conditions in cyanobacteria, but severe stress conditions might repress the translation activity. It should also be noted that the effect of salt stress on translation activity may vary depending on the strain of cyanobacterium.

In the case of downregulated DEGs, 99 genes were downregulated under both stresses (Figure 4B). In particular, 86% of 115 DEGs whose expression was downregulated under salt stress were also downregulated under antibiotic stress.

Up- and Down-Regulated DEGs under Stress Conditions

Table 1 and Table 2 list the genes whose expression was remarkably induced or repressed by both salt stress and antibiotic stress, respectively.

Of the genes that were upregulated by both salt stress and antibiotic stress, 57 out of 119 genes were genes encoding hypothetical proteins. Among the functionally annotated genes, *Synpcc7942_2510*, which encodes bacterial translation initiation factor 1 (bIF-1), was found to be directly involved in protein translation. Two genes, *Synpcc7942_0445* and *Synpcc7942_1126*, encoding ABC-type membrane proteins are thought to be responsible for the regulation of cell permeability to both salt stress and antibiotic stress. The genes coding thioredoxin, *Synpcc7942_1793*, and thioredoxin peroxidase, *Synpcc7942_2309*, might be involved in oxidative stress defense against ROS generated by salt and antibiotic stress. In addition, investigation of the target genes of the factors involved in transcriptional regulation encoded by genes such as *Synpcc7942_0938*, *Synpcc7942_1291*, *Synpcc7942_1310*, *Synpcc7942_1693*, *Synpcc7942_ 2356*, and *Synpcc7942_2416* would be an interesting subject for future study.

Table 1. Genes whose expression was upregulated by both salt and antibiotic stresses on day 3 after stress treatments

Gene ID	Log2FC				Product
	0.4 M NaCl		50 µg/ml streptomycin		
	Day 1	Day 3	Day 1	Day 3	
Synpcc7942_0064	3.9	2.1	3.0	2.1	hypothetical protein
Synpcc7942_0074	-*	-*	-*	-*	conserved hypothetical protein
Synpcc7942_0140	3.5	3.3	2.4	3.3	conserved hypothetical protein
Synpcc7942_0159	3.5	3.1	3.4	3.1	Mov34/MPN/PAD-1
Synpcc7942_0160	4.4	4.4	2.8	4.4	GTP-binding protein Era
Synpcc7942_0191	3.3	3.4	2.9	3.4	Serine--glyoxylate transaminase
Synpcc7942_0206	2.6	5.3	4.3	5.3	Nicotinic acid phosphoribosyltransferase-like
Synpcc7942_0207	3.9	4.9	2.2	4.9	hypothetical protein
Synpcc7942_0219	3.0	3.0	5.1	3.0	conserved hypothetical protein
Synpcc7942_0229	6.2	6.8	6.5	6.8	hypothetical protein
Synpcc7942_0373	6.0	6.2	4.4	6.2	conserved hypothetical protein
Synpcc7942_0374	6.5	6.5	7.1	6.5	hypothetical protein
Synpcc7942_0375	2.8	3.3	3.8	3.3	processing protease
Synpcc7942_0400	3.7	2.9	2.4	2.9	anthranilate synthase, component II
Synpcc7942_0406	4.2	4.1	2.9	4.1	conserved hypothetical protein
Synpcc7942_0408	2.4	5.5	2.5	5.5	phosphoribosylanthranilate isomerase
Synpcc7942_0443	5.2	4.3	3.6	4.3	conserved hypothetical protein
Synpcc7942_0444	5.9	4.7	5.0	4.7	conserved hypothetical protein
Synpcc7942_0445	2.5	2.2	4.2	2.2	ABC-type dipeptide/oligopeptide/nickel transport systems permease components-like
Synpcc7942_0447	2.4	2.7	2.4	2.7	DNA-(apurinic or apyrimidinic site) lyase/endonuclease III
Synpcc7942_0455	3.7	3.9	3.1	3.9	queuine tRNA-ribosyltransferase
Synpcc7942_0467	3.6	3.4	3.5	3.4	Peptidase, metallopeptidases
Synpcc7942_0468	3.2	2.4	2.9	2.4	preprotein translocase, SecG subunit
Synpcc7942_0510	3.3	2.5	2.5	2.5	2-hydroxy-3-keto-5-methylthiopentenyl-1-phosphate phosphatase
Synpcc7942_0542	3.5	3.1	3.0	3.1	lipoic acid synthetase
Synpcc7942_0553	3.9	3.8	2.7	3.8	Secretion protein HlyD
Synpcc7942_0555	4.4	4.0	3.6	4.0	zinc metalloprotease
Synpcc7942_0561	3.9	3.7	3.5	3.7	hypothetical protein
Synpcc7942_0562	4.5	3.7	2.5	3.7	conserved hypothetical protein
Synpcc7942_0584	5.2	6.3	5.7	6.3	ribose-5-phosphate isomerase
Synpcc7942_0585	4.3	3.9	3.9	3.9	PDZ/DHR/GLGF
Synpcc7942_0590	2.3	2.6	2.4	2.6	membrane protein-like
Synpcc7942_0591	2.6	2.6	2.5	2.6	membrane protein-like
Synpcc7942_0597	2.9	2.8	2.5	2.8	1,4-Dihydroxy-2-naphthoate synthase
Synpcc7942_0665	3.7	3.0	2.8	3.0	conserved hypothetical protein
Synpcc7942_0754	3.3	2.8	4.3	2.8	hypothetical protein
Synpcc7942_0755	3.6	3.9	3.0	3.9	conserved hypothetical protein
Synpcc7942_0786	3.2	2.4	3.0	2.4	conserved hypothetical protein
Synpcc7942_0820	3.0	2.4	3.2	2.4	conserved hypothetical protein
Synpcc7942_0864	3.5	3.5	4.2	3.5	conserved hypothetical protein

Table 1. (Continued)

Gene ID	Log2FC				Product
	0.4 M NaCl		50 μg/ml streptomycin		
	Day 1	Day 3	Day 1	Day 3	
Synpcc7942_0865	3.5	3.6	3.2	3.6	conserved hypothetical protein
Synpcc7942_0867	4.4	4.2	4.6	4.2	hypothetical protein
Synpcc7942_0889	4.0	4.6	2.5	4.6	conserved hypothetical protein
Synpcc7942_0914	5.3	4.1	3.4	4.1	tRNA-(MS(2)IO(6)A)-hydroxylase-like
Synpcc7942_0937	3.9	4.7	4.8	4.7	conserved hypothetical protein
Synpcc7942_0938	4.4	6.8	5.4	6.8	transcriptional regulator, ArsR family
Synpcc7942_0939	3.2	4.4	3.9	4.4	adenylylsulfate kinase
Synpcc7942_0964	4.3	3.4	2.8	3.4	conserved hypothetical protein
Synpcc7942_0980	2.7	4.6	3.9	4.6	conserved hypothetical protein
Synpcc7942_1126	3.4	3.5	3.8	3.5	ABC transporter permease protein
Synpcc7942_1193	4.9	3.7	4.2	3.7	Phospholipid/glycerol acyltransferase
Synpcc7942_1204	3.4	4.9	3.9	4.9	Prevent-host-death protein
Synpcc7942_1260	3.9	3.8	4.8	3.8	conserved hypothetical protein
Synpcc7942_1290	2.7	5.1	4.2	5.1	SmtA metallothionein
Synpcc7942_1291	3.4	2.5	4.1	2.5	transcriptional regulator, ArsR family
Synpcc7942_1310	4.2	2.7	2.6	2.7	ndhF3 operon transcriptional regulator
Synpcc7942_1327	5.3	4.6	4.3	4.6	hypothetical protein
Synpcc7942_1328	5.7	5.8	4.6	5.8	conserved hypothetical protein
Synpcc7942_1338	3.3	3.5	4.4	3.5	conserved hypothetical protein
Synpcc7942_1362	2.7	3.0	2.7	3.0	conserved hypothetical protein
Synpcc7942_1363	2.7	3.7	2.1	3.7	hypothetical protein
Synpcc7942_1384	4.3	4.3	3.5	4.3	conserved hypothetical protein
Synpcc7942_1405	3.4	4.0	4.6	4.0	conserved hypothetical protein
Synpcc7942_1561	2.9	3.9	4.0	3.9	conserved hypothetical protein
Synpcc7942_1562	5.2	7.0	8.8	7.0	ADP-ribosylglycohydrolase-like
Synpcc7942_1596	2.9	3.8	3.5	3.8	probable short-chain dehydrogenase
Synpcc7942_1602	3.2	3.5	3.8	3.5	RNA methyltransferase TrmH, group 1
Synpcc7942_1603	4.8	3.7	3.7	3.7	beta-lactamase
Synpcc7942_1604	3.9	2.4	3.0	2.4	conserved hypothetical protein
Synpcc7942_1606	9.1	7.2	8.3	7.2	Beta-Ig-H3/fasciclin
Synpcc7942_1613	2.8	2.6	2.2	2.6	conserved hypothetical protein
Synpcc7942_1641	4.6	3.5	5.0	3.5	hypothetical protein
Synpcc7942_1642	3.2	2.9	2.3	2.9	conserved hypothetical protein
Synpcc7942_1649	3.6	4.6	9.0	4.6	rubrerythrin
Synpcc7942_1693	4.4	2.8	4.4	2.8	response regulator receiver domain protein (CheY-like)
Synpcc7942_1710	3.7	3.4	3.3	3.4	RNA polymerase subunit omega (rpoZ)
Synpcc7942_1711	3.6	3.3	2.7	3.3	conserved hypothetical protein
Synpcc7942_1712	2.7	3.0	2.8	3.0	conserved hypothetical protein
Synpcc7942_1731	3.2	3.3	2.6	3.3	conserved hypothetical protein
Synpcc7942_1752	2.4	2.5	2.3	2.5	Armadillo:PBS lyase HEAT-like repeat
Synpcc7942_1761	3.6	3.7	3.4	3.7	conserved hypothetical protein
Synpcc7942_1765	3.6	2.9	2.6	2.9	conserved hypothetical protein
Synpcc7942_1769	3.9	4.9	2.9	4.9	conserved hypothetical protein
Synpcc7942_1791	4.9	5.5	6.4	5.5	conserved hypothetical protein

Gene ID	Log2FC				Product
	0.4 M NaCl		50 µg/ml streptomycin		
	Day 1	Day 3	Day 1	Day 3	
Synpcc7942_1793	3.8	4.0	4.3	4.0	thioredoxin
Synpcc7942_1799	4.4	3.5	5.4	3.5	Hydrogenase expression/formation protein HypE
Synpcc7942_1800	4.4	3.9	5.1	3.9	conserved hypothetical protein
Synpcc7942_1801	5.9	7.8	7.3	7.8	conserved hypothetical protein
Synpcc7942_1812	3.8	3.0	4.1	3.0	hypothetical protein
Synpcc7942_1818	3.0	3.9	2.9	3.9	conserved hypothetical protein
Synpcc7942_1862	3.0	2.6	2.4	2.6	conserved hypothetical protein
Synpcc7942_1870	5.7	6.6	4.4	6.6	Secretion protein HlyD
Synpcc7942_1896	2.4	2.3	3.0	2.3	N-acetylglutamate synthase/ glutamate N-acetyltransferase
Synpcc7942_1923	4.1	3.1	3.5	3.1	sigma-24 (FecI-like)
Synpcc7942_1933	3.2	3.0	4.3	3.0	Isopentenyl diphosphate delta isomerase, FMN-dependent
Synpcc7942_1934	3.2	3.3	3.9	3.3	D-alanyl-D-alanine carboxypeptidase/D-alanyl-D-alanine-endopeptidase
Synpcc7942_1957	2.5	2.5	4.0	2.5	cyclic nucleotide-binding domain (cNMP-BD) protein
Synpcc7942_1960	3.5	3.7	3.0	3.7	conserved hypothetical protein
Synpcc7942_1993	3.3	3.6	4.1	3.6	methylthioribulose-1-phosphate dehydratase
Synpcc7942_2007	3.9	2.8	2.4	2.8	Phosphopantetheine adenylyltransferase
Synpcc7942_2013	3.2	4.4	2.8	4.4	conserved hypothetical protein
Synpcc7942_2040	3.9	3.5	4.4	3.5	sugar fermentation stimulation protein
Synpcc7942_2088	2.1	3.6	4.2	3.6	conserved hypothetical protein
Synpcc7942_2123	3.0	3.3	2.7	3.3	anthranilate phosphoribosyltransferase
Synpcc7942_2124	4.1	6.4	4.5	6.4	acetylpolyamine aminohydolase
Synpcc7942_2134	3.8	3.3	3.4	3.3	Cell wall hydrolase/autolysin
Synpcc7942_2256	3.1	2.4	3.7	2.4	conserved hypothetical protein
Synpcc7942_2260	2.4	2.7	2.8	2.7	conserved hypothetical protein
Synpcc7942_2309	3.2	3.7	6.1	3.7	thioredoxin peroxidase
Synpcc7942_2356	4.1	5.0	6.9	5.0	transcriptional regulator, Crp/Fnr family
Synpcc7942_2357	2.5	2.3	3.3	2.3	conserved hypothetical protein
Synpcc7942_2414	4.6	5.9	4.9	5.9	conserved hypothetical protein
Synpcc7942_2416	2.8	4.0	3.5	4.0	two component transcriptional regulator, winged helix family
Synpcc7942_2419	2.6	2.9	2.7	2.9	conserved hypothetical protein
Synpcc7942_2445	3.1	4.4	5.4	4.4	Phosphate binding protein
Synpcc7942_2458	3.9	3.7	3.3	3.7	ComEC/Rec2-related protein
Synpcc7942_2510	2.5	3.5	3.6	3.5	bacterial translation initiation factor 1 (bIF-1)
Synpcc7942_2514	3.6	4.3	3.9	4.3	ornithine carbamoyltransferase
Synpcc7942_2589	4.7	3.1	2.6	3.1	2-phosphosulfolactate phosphatase

*: Since there was no expression count in control, FC could not be calculated.

Table 2. Genes whose expression was downregulated by both salt and antibiotic stresses for three days after stress treatments

Gene ID	Log2FC				Product
	0.4 M NaCl		50 μg/ml streptomycin		
	Day 1	Day 3	Day 1	Day 3	
Synpcc7942_0053	-1.3	-4.1	-7.1	-4.1	conserved hypothetical protein
Synpcc7942_0054	-1.9	-3.4	-6.3	-3.4	conserved hypothetical protein
Synpcc7942_0055	-2.2	-4.2	-5.0	-4.2	hypothetical protein
Synpcc7942_0061	-2.4	-3.2	-5.4	-3.2	CDP-glucose 4,6-dehydratase
Synpcc7942_0062	-1.5	-1.6	-4.8	-1.6	glucose-1-phosphate cytidylyltransferase
Synpcc7942_0069	-3.3	-4.2	-3.1	-4.2	hypothetical protein
Synpcc7942_0122	-2.3	-5.1	-6.9	-5.1	EAL
Synpcc7942_0180	-2.0	-3.2	-8.5	-3.2	conserved hypothetical protein
Synpcc7942_0181	-1.3	-2.4	-6.3	-2.4	ATPase
Synpcc7942_0267	-2.5	-3.2	-5.3	-3.2	methyltransferase GidB
Synpcc7942_0312	-2.1	-2.8	-2.2	-2.8	Protein of unknown function UPF0102
Synpcc7942_0370	-1.1	-2.8	-2.1	-2.8	O-acetylhomoserine/O-acetylserine sulfhydrylase
Synpcc7942_0381	-1.7	-5.0	-1.6	-5.0	conserved hypothetical protein
Synpcc7942_0490	-1.5	-2.7	-4.0	-2.7	diguanylate cyclase with PAS/PAC sensor
Synpcc7942_0551	-2.3	-4.4	-4.6	-4.4	conserved hypothetical protein
Synpcc7942_0573	-3.9	-4.0	-5.4	-4.0	hypothetical protein
Synpcc7942_0592	-2.2	-3.1	-3.9	-3.1	6-phosphofructokinase
Synpcc7942_0680	-1.8	-3.8	-1.7	-3.8	C-5 sterol desaturase
Synpcc7942_0759	-2.0	-3.8	-6.9	-3.8	hypothetical protein
Synpcc7942_0760	-1.5	-3.9	-7.3	-3.9	hypothetical protein
Synpcc7942_0768	-3.2	-3.6	-3.7	-3.6	conserved hypothetical protein
Synpcc7942_0769	-2.3	-2.5	-3.0	-2.5	conserved hypothetical protein
Synpcc7942_0770	-1.9	-2.2	-3.7	-2.2	conserved hypothetical protein
Synpcc7942_0780	-3.0	-4.4	-4.1	-4.4	serine/threonine protein kinase
Synpcc7942_0781	-2.9	-5.5	-7.1	-5.5	phosphoenolpyruvate synthase
Synpcc7942_0834	-4.2	-6.3	-6.4	-6.3	conserved hypothetical protein
Synpcc7942_0840	-1.8	-6.1	-7.1	-6.1	conserved hypothetical protein
Synpcc7942_0855	-2.9	-3.1	-8.3	-3.1	response regulator receiver domain protein (CheY-like)
Synpcc7942_0856	-3.2	-3.7	-8.3	-3.7	response regulator receiver domain protein (CheY-like)
Synpcc7942_0857	-3.0	-3.4	-5.8	-3.4	CheW protein
Synpcc7942_0858	-1.8	-2.8	-8.3	-2.8	methyl-accepting chemotaxis sensory transducer with phytochrome sensor
Synpcc7942_0859	-2.1	-3.3	-7.5	-3.3	CheA signal transduction histidine kinase
Synpcc7942_0860	-1.9	-3.3	-7.6	-3.3	CheW protein
Synpcc7942_0905	-3.1	-3.5	-5.4	-3.5	hypothetical protein
Synpcc7942_0906	-1.2	-1.5	-3.1	-1.5	hypothetical protein
Synpcc7942_1040	-2.5	-2.9	-2.2	-2.9	conserved hypothetical protein
Synpcc7942_1095	-1.3	-1.6	-2.2	-1.6	conserved hypothetical protein
Synpcc7942_1117	-1.6	-4.1	-3.0	-4.1	conserved hypothetical protein
Synpcc7942_1133	-1.1	-3.3	-5.0	-3.3	bacteriocin-processing peptidase. Cysteine peptidase. MEROPS family C39

Gene ID	Log2FC				Product
	0.4 M NaCl		50 µg/ml streptomycin		
	Day 1	Day 3	Day 1	Day 3	
Synpcc7942_1134	-2.0	-4.3	-6.5	-4.3	hypothetical protein
Synpcc7942_1150	-1.6	-2.0	-1.3	-2.0	conserved hypothetical protein
Synpcc7942_1153	-2.4	-2.0	-3.0	-2.0	conserved hypothetical protein
Synpcc7942_1249	-1.5	-2.2	-4.5	-2.2	photosystem I reactionCenter subunit IX
Synpcc7942_1345	-1.2	-4.3	-3.9	-4.3	NADH dehydrogenase subunit 6
Synpcc7942_1346	-1.5	-4.4	-4.4	-4.4	NADH dehydrogenase subunit 4L
Synpcc7942_1427	-2.0	-5.5	-5.7	-5.5	ribulose 1,5-bisphosphate carboxylase small subunit
Synpcc7942_1469	-1.1	-1.4	-5.9	-1.4	putative multicomponent Na+:H+ antiporter subunit B
Synpcc7942_1470	-1.6	-1.5	-6.5	-1.5	conserved hypothetical protein
Synpcc7942_1471	-1.6	-1.5	-6.6	-1.5	conserved hypothetical protein
Synpcc7942_1472	-1.5	-1.6	-6.3	-1.6	conserved hypothetical protein
Synpcc7942_1473	-1.7	-1.8	-5.6	-1.8	putative multicomponent Na+:H+ antiporter subunit D
Synpcc7942_1475	-8.0	-8.9	-6.3	-8.9	sodium-dependent bicarbonate transporter
Synpcc7942_1476	-1.6	-1.7	-3.7	-1.7	conserved hypothetical protein
Synpcc7942_1488	-7.7	-7.9	-6.7	-7.9	membrane protein
Synpcc7942_1489	-5.9	-7.5	-6.1	-7.5	nitrate transport permease
Synpcc7942_1490	-5.4	-5.7	-5.9	-5.7	nitrate transport ATP-binding subunits C and D
Synpcc7942_1491	-4.5	-5.7	-3.0	-5.7	nitrate transport ATP-binding subunits C and D
Synpcc7942_1504	-1.7	-3.1	-5.4	-3.1	conserved hypothetical protein
Synpcc7942_1510	-2.6	-3.2	-5.1	-3.2	group3 RNA polymerase sigma factor SigF
Synpcc7942_1572	-2.5	-3.5	-5.4	-3.5	dehydrogenase subunit-like protein
Synpcc7942_1573	-2.8	-4.0	-6.9	-4.0	3-oxoacyl-(acyl-carrier protein) reductase
Synpcc7942_1574	-1.9	-2.7	-4.5	-2.7	probable glucosidase
Synpcc7942_1587	-2.3	-4.0	-4.7	-4.0	integral membrane protein-like
Synpcc7942_1607	-1.7	-1.8	-2.4	-1.8	probable porin%3B major outer membrane protein
Synpcc7942_1633	-2.4	-4.2	-6.2	-4.2	conserved hypothetical protein
Synpcc7942_1655	-1.4	-3.7	-5.1	-3.7	hypothetical protein
Synpcc7942_1661	-5.5	-6.6	-4.9	-6.6	conserved hypothetical protein
Synpcc7942_1724	-3.1	-4.2	-2.8	-4.2	conserved hypothetical protein
Synpcc7942_1798	-1.3	-3.9	-2.6	-3.9	conserved hypothetical protein
Synpcc7942_1811	-1.7	-3.0	-5.9	-3.0	diguanylate cyclase (GGDEF domain) with PAS/PAC sensor
Synpcc7942_2048	-2.0	-5.8	-6.4	-5.8	photosystem I core protein PsaB
Synpcc7942_2049	-1.9	-5.1	-4.3	-5.1	photosystem I core protein PsaA
Synpcc7942_2094	-2.0	-5.0	-6.3	-5.0	Beta-Ig-H3/fasciclin
Synpcc7942_2096	-3.5	-3.6	-5.3	-3.6	diguanylate cyclase with GAF sensor
Synpcc7942_2151	-2.5	-3.6	-6.5	-3.6	Cellulose synthase (UDP-forming)
Synpcc7942_2152	-2.1	-2.6	-4.7	-2.6	conserved hypothetical protein
Synpcc7942_2175	-1.9	-4.6	-5.5	-4.6	transport system substrate-binding protein
Synpcc7942_2196	-1.7	-2.4	-1.9	-2.4	conserved hypothetical protein
Synpcc7942_2306	-3.4	-6.4	-6.3	-6.4	Heat shock protein DnaJ-like
Synpcc7942_2307	-3.4	-7.3	-7.3	-7.3	conserved hypothetical protein
Synpcc7942_2308	-2.8	-4.4	-4.9	-4.4	aminomethyltransferase

Table 2. (Continued)

Gene ID	Log2FC				Product
	0.4 M NaCl		50 µg/ml streptomycin		
	Day 1	Day 3	Day 1	Day 3	
Synpcc7942_2349	-1.6	-1.7	-1.2	-1.7	twitching motility protein
Synpcc7942_2352	-2.0	-5.4	-7.1	-5.4	sigma 54 modulation protein/ SSU ribosomal protein S30P
Synpcc7942_2384	-2.9	-3.2	-5.3	-3.2	Pyruvate:ferredoxin (flavodoxin) oxidoreductase
Synpcc7942_2385	-2.9	-3.3	-5.4	-3.3	conserved hypothetical protein
Synpcc7942_2386	-1.9	-2.3	-3.6	-2.3	hydrogenase expression/formation protein HypD
Synpcc7942_2387	-2.2	-2.5	-3.7	-2.5	conserved hypothetical protein
Synpcc7942_2388	-3.1	-3.0	-5.3	-3.0	Oxalate decarboxylase
Synpcc7942_2391	-3.3	-4.3	-5.4	-4.3	oxidoreductase-like
Synpcc7942_2451	-1.3	-3.5	-3.3	-3.5	putative type IV pilus assembly protein PilO
Synpcc7942_2452	-1.4	-3.4	-4.1	-3.4	Tfp pilus assembly protein PilN-like
Synpcc7942_2485	-1.8	-1.7	-6.6	-1.7	hypothetical protein
Synpcc7942_2498	-1.5	-2.8	-4.5	-2.8	Band 7 protein
Synpcc7942_2499	-1.6	-3.2	-5.1	-3.2	Band 7 protein
Synpcc7942_2519	-1.9	-2.5	-5.5	-2.5	diguanylate cyclase/phosphodiesterase
Synpcc7942_2535	-2.5	-3.9	-5.6	-3.9	diguanylate cyclase (GGDEF domain) with PAS/PAC and GAF sensors
Synpcc7942_2604	-1.1	-1.7	-6.6	-1.7	cytochrome c oxidase subunit III
Synpcc7942_2609	-1.4	-1.2	-3.7	-1.2	conserved hypothetical protein
Synpcc7942_B2646	-2.3	-3.9	-2.3	-3.9	two-component sensor histidine kinase

Of the downregulated genes, 40 out of 99 genes encoded hypothetical proteins. Among the functionally annotated genes, suppression of some genes involved in major metabolic pathways and photosynthesis was observed. Regarding the primary metabolic pathway, the expression of several genes (*Synpccc7942_0592*, *Synpccc7942_0781*, and *Synpccc7942_2384*) involved in glycolysis was decreased. For photosynthesis, the expression of genes encoding components of photosystem I (*Synpccc7942_1249*, *Synpccc7942_2048*, and *Synpccc7942_2049*), cytochrome c oxidase subunit III (*Synpccc 7942_2604*), and ribulose 1,5-bisphosphate carboxylase small subunit (*Synpccc7942_1427*) were remarkably suppressed. The expression of genes, *Synpcc7942_0855-0860* and *Synpcc7942_B2646*, encoding factors involved in the two-component regulatory system was also decreased significantly.

Conclusion

In this study, freshwater model cyanobacteria were treated with salt stress and antibiotic stress at concentrations that caused severe growth inhibition. Transcriptome analysis showed the genes whose expression changed remarkably under both stresses. These genes are thought to be responsible for a common molecular mechanism that responds to both stresses. Although the relationship between individual gene functions and stress response mechanisms has not yet been clarified, the information provided in this chapter may contribute to the understanding of abiotic stress tolerance mechanisms in cyanobacteria.

Acknowledgment

This work was supported in part by the research support fund of the Research Institute of Meijo University, FY2020 Research Area Designated Support Project Expenses (to H.K.).

References

Billis, K., Billini, M., Tripp, H. J., Kyrpides, N. C., and Mavromatis, K. (2014). Comparative transcriptomics between *Synechococcus* PCC 7942 and *Synechocystis* PCC 6803 provide insights into mechanisms of stress acclimation. *PLoS One*, 9, e109738.

Dong, Z., Sun, T., Zhang, W., and Chen, L. (2023). Inproved salt tolernace of *Synechococcus elongatus* PCC 7942 by heterologous synthesis of compatible solute ectoine. *Frontiers in Microbiology*, 14, 1123081.

Du, Y., Wang, J., Zhu, F., Mai, D., Xiang, Z., Chen, J., and Guo, R. (2018). Comprehensive assessment of three typical antibiotics on cyanobacteria (*Microcystis aeruginosa*): The impact and recovery capability. *Ecotoxicol Environ Saf*, 160, 84-93.

Garcia-Pichel, F., Nübel, U., and Muyzer, G. (1998). The phylogeny of unicellular, extremely halotolerant cyanobacteria. *Arch Microbiol*, 169, 469-482.

Häder, D. P., Kumar, H. D., Smith, R. C., and Worrest, R. C. (2007). Effects of solar UV radiation on aquatic ecosystems and interactions with climate change. *Photochem Photobiol Sci*, 6, 267-285.

Hagemann, M. (2011). Molecular biology of cyanobacterial salt acclimation. *FEMS Microbiol Rev*, 35, 87-123.

Kageyama, H., Tripathi, K., Rai, A. K., Cha-Um, S., Waditee-Sirisattha, R., and Takabe, T. (2011). An alkaline phosphatase/phosphodiesterase, PhoD, induced by salt stress and secreted out of the cells of *Aphanothece halophytica*, a halotolerant cyanobacterium. *Appl Environ Microbiol*, 77, 5178-5183.

Kanesaki, Y., Suzuki, I., Allakhverdiev, S. I., Mikami, K., and Murata, N. (2002). Salt stress and hyperosmotic stress regulate the expression of different sets of genes in *Synechocystis* sp. PCC 6803. *Biochem Biophys Res Commun*, 290, 339-348.

Mataloni, G., and Komárek, J. (2004). *Gloeocapsopsis aurea*, a new subaerophytic cyanobacterium from maritime Antarctica. *Polar Biology*, 27, 623-628.

Patipong, T., Hibino, T., Waditee-Sirisattha, R., and Kageyama, H. (2019a). Induction of antioxidative activity and antioxidant molecules in the halotolerant cyanobacterium *Halothece* sp. PCC7418 by temperature shift. *Natural Product Communications*, 14, 1-6.

Patipong, T., Ngoennet, S., Honda, M., Hibino, T., Waditee-Sirisattha, R., and Kageyama, H. (2019b). A class I fructose-1,6-bisphosphate aldolase is associated with salt stress tolerance in a halotolerant cyanobacterium *Halothece* sp. PCC 7418. *Arch Biochem Biophys*, 672. 108059.

Qian, H., Li, J., Pan, X., Sun, Z., Ye, C., Jin, G., and Fu, Z. (2012). Effects of streptomycin on growth of algae *Chlorella vulgaris* and *Microcystis aeruginosa. Environ Toxicol*, 27, 229-237.

Sharma, D., Cukras, A. R., Rogers, E. J., Southworth, D. R., and Green, R. (2007). Mutational analysis of S12 protein and implications for the accuracy of decoding by the ribosome. *J Mol Biol*, 374, 1065-1076.

Viveiros, M., Dupont, M., Rodrigues, L., Couto, I., Davin-Regli, A., Martins, M., Pagès, J. M., and Amaral, L. (2007). Antibiotic stress, genetic response and altered permeability of *E. coli*. *PLoS One*, 2, e365.

Waditee-Sirisattha, R., Ito, H., and Kageyama, H. (2022). Global transcriptional and circadian regulation in a halotolerant cyanobacterium *Halothece* sp. PCC7418. *Sci Rep*, 12, 13190,

Waditee-Sirisattha, R., and Kageyama, H. (2022a). Extremophilic cyanobacteria. In: *Cyanobacterial Physiology: From Fundamentals to Biotechnology*, Kageyama H., Waditee-Sirisattha R. (eds.) pp. 85-99. Academic Press (An imprint of Elsevier): San Diego, CA, USA.

Waditee-Sirisattha, R., and Kageyama, H. (2022b). Global transcriptome analyses and regulatory mechanisms in *Halothece* sp. PCC 7418 exposed to abiotic stresses. *Appl Microbiol Biotechnol*, 106, 6641-6655.

Waditee-Sirisattha, R., and Kageyama, H. (2023). Halotolerance, stress mechanisms, and circadian clock of salt-tolerant cyanobacteria. *Appl Microbiol Biotechnol*, 107, 1129-1141.

Waditee-Sirisattha, R., Kageyama, H., Fukaya, M., Rai, V., and Takabe, T. (2015). Nitrate and amino acid availability affects glycine betaine and mycosporine-2-glycine in response to changes of salinity in a halotolerant cyanobacterium *Aphanothece halophytica*. *FEMS Microbiol Lett*, 362, fnv198.

Waditee-Sirisattha, R., Kageyama, H., Sopun, W., Tanaka, Y., and Takabe, T. (2014). Identification and upregulation of biosynthetic genes required for accumulation of Mycosporine-2-glycine under salt stress conditions in the halotolerant cyanobacterium *Aphanothece halophytica*. *Appl Environ Microbiol*, 80, 1763-1769.

Waditee, R., Bhuiyan, M. N., Rai, V., Aoki, K., Tanaka, Y., Hibino, T., Suzuki, S., Takano, J., Jagendorf, A. T., Takabe, T., et al. (2005). Genes for direct methylation of glycine provide high levels of glycinebetaine and abiotic-stress tolerance in *Synechococcus* and *Arabidopsis*. *Proc Natl Acad Sci U S A*, 102, 1318-1323.

Waditee, R., Tanaka, Y., Aoki, K., Hibino, T., Jikuya, H., Takano, J., Takabe, T., and Takabe, T. (2003). Isolation and functional characterization of *N*-methyltransferases that catalyze betaine synthesis from glycine in a halotolerant photosynthetic organism *Aphanothece halophytica*. *J Biol Chem*, 278, 4932-4942.

Whitton, B. A., and Potts, M. (2000). Introduction of cyanobacteria. In: *The Ecology of Cyanobacteria: Their Deiversity in Time and Space*, Whitton B.A., Potts M. (eds.) pp. 1-10. Kluwer Academic Publishers: Norwell, MA, USA.

Xiong, W., Lee, T. C., Rommelfanger, S., Gjersing, E., Cano, M., Maness, P. C., Ghirardi, M., and Yu, J. (2015). Phosphoketolase pathway contributes to carbon metabolism in cyanobacteria. *Nat Plants*, 2, 15187

Yalcin, Y. S., Aydin, B. N., Sayadujjhara, M., and Sitther, V. (2022). Antibiotic-induced changes in pigment accumulation, photosystem II, and membrane permeability in a model cyanobacterium. *Front Microbiol*, 13, 930357.

Chapter 4

The Characterization and Biological Activity of Plant Lectins: A Review

Ludmila Maria Gonçalves Godoi de Camargos, MS
Débora de Oliveira Lopes, PhD
and Ana Hortência Fonsêca Castro*, PhD
Universidade Federal de São João del-Rei, Divinópolis, Minas Gerais, Brazil

Abstract

Lectins are proteins or glycoproteins of a non-immune nature, capable of reversibly binding to specific carbohydrates without causing changes in their structure. These proteins are known to be distributed among all living organisms and directly participate in several biological processes, among them cell recognition, inflammation, signaling of many metabolic pathways, viral and bacterial pathogenesis. Lectins also have an ability to agglutinate erythrocytes in vitro and are also associated with agglutinins or phytohemagglutinins. In plants, studies reveal that these proteins act in the protection and defense against pathogens and predators, as transporters of plant hormones and in symbiotic plant-microorganism interactions. The Leguminosae family represents the most investigated and well-described group of lectins within the plant kingdom, mainly due to its wide diversity and specificity. However, obtaining proteins directly from plants is not always feasible since it includes expensive, laborious, and low-yield steps. Thus, biotechnological techniques, as well as several expression systems (prokaryotes and eukaryotes), have been proposed for heterologous production of lectins of agronomic and biomedical interest. Chromatographic assays have aided in the isolation and purification of

* Corresponding Author's Email: acastro@ufsj.edu.br.

In: Advances in Biology. Volume 4
Editor: Charles D. Grant
ISBN: 979-8-88697-999-2

heterologous proteins for application in histochemical, biochemical, and medicinal studies. The biological activities of plant lectins are diverse, and *in vivo*, they play an important role in the defense mechanisms of plants, mainly due to their multivalence in the recognition and interaction with glycoconjugates of other organisms. *In vitro* studies have demonstrated various medicinal activities of lectins such as antifungal, antitumor, anticarcinogenic, antiviral, and antibacterial effects among other remarkable properties of lectins. This chapter will present a description of plant lectins, in relation to structure, localization, heterologous expression, detection and purification methods, biological activity, and biotechnological applications.

Keywords: recombinant proteins, heterologous expression, Fabaceae, plant lectins, medicinal activities

Introduction

Historically, it is believed that the first lectin was described by the microbiologist Peter Hermann Stillmark in 1888. In his experiments with castor bean seeds (*Ricinus communis*), Stillmark isolated a highly toxic protein and, testing the behavior of this macromolecule (called ricin) in the presence of blood cells, observed their ability to agglutinate erythrocytes *in vitro* (Van Damme, 2014). Later, this phenomenon was studied in more detail by researcher H. Hellin. In his works, Hellin also identified a toxic hemagglutinin in rosary pea (*Abrus precatorius*) extracts, which he designated abrin. In the 1890s, both hemagglutinins became commercially available, allowing the bacteriological biologist Paul Ehrlich to develop studies using them as antigens and establish fundamental basic concepts of immunology. In 1978, ricin became popularly known, and there are reports of its use – without success – in tests for the development of biological weapons during the First and Second World War (Sharon and Lis, 2004).

The use of the term "lectin" was registered for the first time in 1954 to define this class of vegetable proteins of non-immune origin, capable of binding reversibly to specific carbohydrates without causing changes in their structure (Boyd and Shapleigh, 1954). Despite having a wide distribution among different groups of living organisms, lectins are also called agglutinins or phytohemagglutinins since they were originally isolated from plant extracts (Hajtó et al., 2005). Lectins generally show high stability over wide pH and temperature ranges, as well as a high amount of hydroxy amino acids and acidic amino acids, when compared to a low content of sulfur-containing amino acids. In their biosynthesis, most of them undergo post-translational processes, being glycosylated and presenting variable carbohydrate side chains. Lectins may also depend on the presence of Ca^{2+} and Mn^{2+} ions, in specific sites, to perform their hemagglutinating activity and, for this reason, are classified as metalloproteins (Cavada et al., 2020; Peumans and Van Damme, 1995).

Currently, it is known that lectins represent a group of proteins and/or glycoproteins widely distributed among all living organisms and that they participate directly in several biological processes, including cell recognition, inflammation, signaling of many metabolic pathways, and viral and bacterial pathogenesis. However, because they represent biological molecules with high expression rates in plants, most of the lectins studied belong to plant species, which are found in leaves, roots, rhizomes, bulbs, flowers, fruits, tubers, bark, stems, and, predominantly, in seeds and reserve organs (Pietrzyk-Brzezinska and Bujacz, 2020).

Studies show that, in addition to interacting with terminal and subterminal residues of carbohydrates, plant lectins are also capable of interacting with other potentiating molecules, such as secondary metabolites, plant hormones, adenines, porphyrins, and antibiotics (Cavada et al., 2020).

Although the physiological role of lectins still needs to be better elucidated, studies have revealed that these proteins act in protection and defense against pathogens and predators, as transporters of plant hormones and in plant-microorganism symbiotic interactions. Most often, the performance of these functions involves the interaction between a carbohydrate-recognition domain (CRD) and carbohydrates/ glycoconjugates present on the cell surface. In addition, such interactions occur through non-covalent bonds such as hydrogen bonds, Van der Waals, and hydrophobic forces (Cavada et al., 2020; Camacho, 2007).

Lectins form a heterogeneous group since the proteins that compose it are oligomeric in nature and vary in size, molecular organization, CRD type, and physical and biochemical properties. According to Hajtó et al. (2005), lectins are classified according to their specificity of binding to sugars (glucose and mannose, mannose, D-galactose, N-acetylglucosamine, and sialic acid) (Figure 1) and are related to the monosaccharide that causes the greatest inhibition of erythrocyte agglutination induced by the lectin or precipitation of carbohydrate-containing polymers. However, this classification based on binding affinity to mono- and oligosaccharides varies widely within the literature.

Based on their molecular structure, lectins are classified into merolectins, hololectins, chimerolectins, and superlectins (Figure 2). Merolectins are proteins that have only one carbohydrate-binding domain and are unable to precipitate glycoconjugates and agglutinate cells due to their monovalent nature. Hololectins are the most abundant among plants and have two or more binding sites, that are identical or homologous. They are capable of causing the precipitation of glycoconjugates and the agglutination of cells due to their bi- or multivalent nature. On the other hand, chimerolectins are proteins that have one or more CRD and another unrelated site, which acts independently and has well-defined catalytic/biological activity. Chimerolectins can behave like merolectins or hololectins, depending on the number of CDRs present in their structure. Superlectins, on the other hand, exhibit at least two binding domains in their structures, capable of recognizing structurally and functionally different carbohydrates (Tsaneva and Van Damme, 2020; Van Damme et al., 1998; Peumans and Van Damme, 1995).

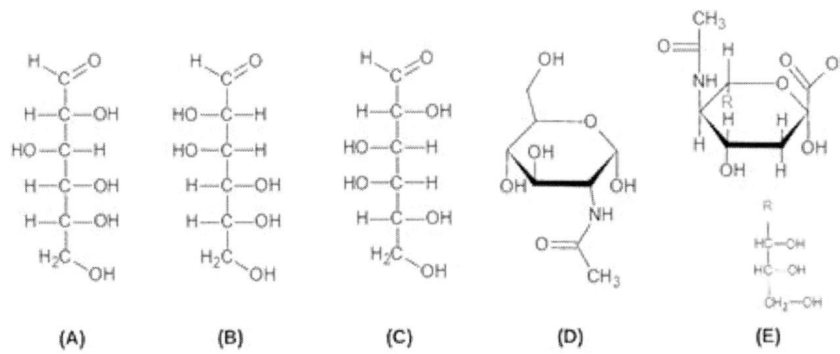

Figure 1. Major sugars commonly binding to lectins - (A) D-glucose; (B) D-mannose; (C) D-galactose, (D) N-acetylglucosamine and (E) sialic acid.

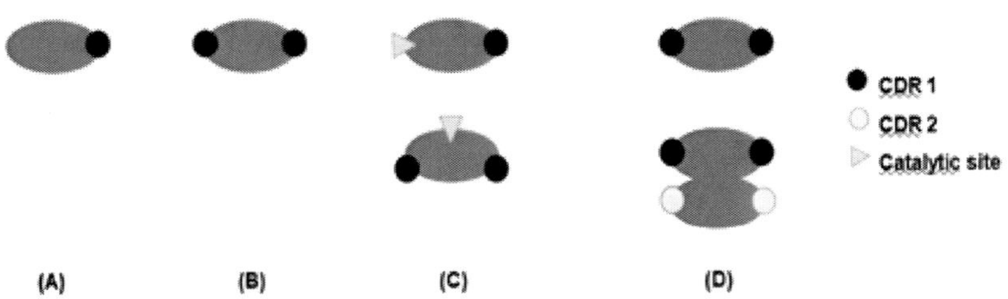

Figure 2. Schematic representation of lectins - (A) Merolectins; (B) Hololectins; (C) Chimerolectins and (D) Superlectins (Adapted from Tsaneva and Van Damme, 2020).

Plant Lectins

Lectins are widely distributed among plant species, having been identified in specimens from the Leguminosae (Fabaceae), Gramineae (Poaceae), Euphorbiaceae families among others. The Leguminosae family presents the most investigated and well-described group of lectins within the plant kingdom, mainly due to its wide diversity and specificity. Thus, several Leguminosae lectins have already been isolated (mainly from mature seeds), purified, and made available for application as bioactive molecules in studies in the agronomic, industrial, medical, and biotechnological areas. The Leguminosae family is classically subdivided into three subfamilies (Caesalpinioideae, Mimosoideae, and Papilionoideae), with emphasis on the latter, which has the largest number of published studies in relation to its lectins (Cavada et al., 2020; Van Damme et al., 1998).

Most plant lectins are commonly isolated from the cotyledon of mature seeds, accounting for 1 to 10% of their total protein content. Since they are proteins, the metabolic pathway for biosynthesis of lectins is common to other excretion and/or storage proteins in intracellular vesicles produced by the plant. The biosynthesis of such proteins occurs concomitantly with the seed maturation process and involves post-translational processing to obtain the functional protein (Van Damme et al., 1998; Peumans and Van Damme, 1995).

The first lectin to be isolated and characterized was concanavalin A (ConA), in 1936, from the common jack bean (*Canavalia ensiformis*), a legume of the Papilionoideae subfamily, Diocleinae subtribe, and *Canavalia* genus. ConA remains the best characterized plant lectin to date (Pietrzyk-Brzezinska and Bujacz, 2020, Van Damme et al, 1998).

Structurally, many plant lectins are considered metalloproteins because they require the presence of divalent cations (Ca^{2+} and Mn^{2+}), in specific binding sites and distinct from the binding domain, for recognition and binding to carbohydrates. Studies demonstrated that such domains are highly conserved in relation to their amino acid chain (Cavada et al., 2020; Van Damme et al., 1998).

In Silico Characterization of Plant Lectins

Advances in the fields of biotechnology and molecular biology have allowed the mass sequencing of the genome of several organisms, as well as increasing the possibility of its better understanding. Thus, the association of mathematical concepts and computer tools is increasingly used to help elucidate the knowledge behind these genomic and proteomic data (Silva, Gonçalves and Lopes, 2020).

Once the genome of a plant species is sequenced, bioinformatics can then be used for *in vitro* characterization of its encoded proteins, including lectins. There are several computer programs and software that allow the alignment of nucleotide and/or protein sequences to search for homologous sequences, to perform phylogenetic analyzes and to predict important physicochemical properties such as hydropathicity, isoelectric point, molecular weight, functional domains, cell location, structures tertiary, etc. (Bonnardel, 2021).

In silico characterization of lectins not only provides greater depth in biological investigations but also allows better planning of *in vivo* studies.

Heterologous Expression of Recombinant Lectins

Obtaining proteins directly from plants is not always feasible since they usually include expensive, laborious, and low-yield steps. Besides that, the functional characterization and detailed study of many lectins can be compromised by the quantity and purity of the protein obtained through the plant extraction process. As a rule, when extracted from plant extracts, what is obtained is a heterogeneous mixture of different lectin isoforms. Several factors are involved in this heterogeneity, including imprecise cutting of the C-terminals of mature polypeptides, multigenic coding, selective expression of these proteins, and the presence of allelic variation in lectin sequences between plant strains. Therefore, in many cases, conventional purification techniques allow the proposition of functional properties and biological activities for a pool of proteins, instead of a specific one (Raemaekers et al., 1999).

The expression of lectins in heterologous systems allows us to explore the sequence-function relationships, through mutagenesis in specific sites, since it enables the production of proteins with a defined amino acid sequence. Thus, cloning techniques for genes of interest, as well as several expression systems (prokaryotes and eukaryotes), have been proposed for the heterologous production of proteins of agronomic, medicinal, or biotechnological interest. Bacteria, fungi, insects, and other animal cells, as well as plants, have been optimized as platforms for the production of recombinant proteins (Islam et al., 2020; Azzoni, 2002, Raemaekers et al., 1999).

The heterologous expression of lectins through these systems is essential for the individualized study of their molecular structure, potential biological activities, and the relationship between these well-defined sequences of amino acids and their functions. However, the ability to detect, express, and purify functional proteins in recombinant systems also represents a major challenge for modern biotechnology. According to Luo, Zhangsun and Tang (2005), bacteria are a promising system for the heterologous expression of non-glycosylated plant lectins and their isoforms. Thus, *Escherichia coli* represents one of the most used expression vectors for obtaining recombinant proteins today, because *E. coli* has characteristics such as fast growth, low cost of cultivation, and it is well characterized proteomics and genetically. However, expression of non-glycosylated lectins in *E. coli* is not always possible. Many forms insoluble inclusion bodies, and in these cases, the lectins must be solubilized by denaturation-renaturation steps, which leads to an inefficient recovery of the active lectin (Raemaekers et al., 1999).

On the other hand, systems designed from yeast, especially *Pichia pastoris*, have been shown to be efficient for the expression of recombinant plant lectins, which require post-translational modifications. In these systems, pre-sequences of the *Saccharomyces cerevisiae* factor α secretion signal are used to ensure the targeting of recombinant proteins to the secretory pathway (Oliveira et al., 2008; Lannoo et al., 2007). However, one of the problems associated with this technique is the presence of critical vacuolar targeting determinants, which can considerably compromise the content of secreted recombinant lectins (Raemaekers et al., 1999).

Transgenic plants also show great potential for the production of recombinant proteins. The advantages of using this expression system are low production cost, reduced risk of contamination for humans, and existence of natural storage organs for the protein (seeds), in addition to well-established harvesting and storage methods. On the other hand, limiting factors such as the complexity of the protocols for transformation/expression of recombinant proteins and low levels of expression obtained represent inherent disadvantages to its use (Islam et al., 2020; Tyurin et al., 2020).

The heterologous expression of plant lectins involves several molecular biology tools and techniques, such as extraction and purification of plant DNA, amplification of the coding sequence of the lectin of interest, cloning in an expression vector, preparation of competent cells (common in systems that use bacteria or yeasts) or agrobacterial transformation of plant species and assays to confirm expression (e.g., Western botting and SDS-PAGE) (Lawanprasert et al., 2020; Tyurin et al., 2020; Camacho, 2007).

Detection and Purification Methods

The ability of lectins to interact reversibly and covalently with free carbohydrates and glycoproteins allows them to be easily detected by agglutination assays. Thus, agglutination techniques are widely used to visualize the agglutinating activity of erythrocytes by lectins. In practice, red blood cells (treated or not) of human origin or of another animal species can be used. Studies have shown that plant lectins have modulated and specific hemagglutinating properties for erythrocytes of different animal origins, as well as selectivity for erythrocytes of a human blood group within the ABO system (Radha and Urla, 2020; Jung et al., 2007; Van Damme et al., 1998). The presence of lectins, when detected by hemagglutination assay, can be confirmed through the phenomenon of hemagglutination inhibition. For this purpose, one or more carbohydrates are added to the test solution, which in the presence of lectins will bind and inhibit the agglutination of red blood cells. At the same time, another way to confirm the presence of lectins in a given sample is using glycoprotein or polysaccharide precipitation assays (Goldstein et al., 2007).

Chromatographic techniques have aided in the isolation and purification of heterologous proteins, as well as in the characterization and analysis of glycoconjugates. For purification of recombinant lectins, affinity chromatography is one of the most widely used assays. In this technique, tails containing histidines (His-tag) are incorporated into the protein structure since their presence facilitates the purification of proteins in a single step and also directly from the cell lysate. The histidine residues present in the tail interact strongly with the metal ion that is part of the chromatographic matrix, allowing the protein to be immobilized and able to be separated from the total protein extract by washing and elution (Puhl, 2008; Chant et al., 2005).

Studies point out that the tails can cause an increase in the solubility and yield of recombinant proteins, in addition to protecting them from attack by intracellular enzymes during the expression process in heterologous systems. On the other hand, there are reports of significant alterations in the biological activities of the target protein, due to the presence of affinity tails in its structure. Since the nature of the tail can positively or negatively affect the properties of the expressed proteins, it is necessary to remove it by cleavage and/or carry out tests that confirm the maintenance of the biological functions of the proteins obtained in their presence (Puhl, 2008; Bucher, Evdokimov and Waugh, 2002).

Biological Activities and Biotechnological Applications

The biological activity of plant lectins has been the subject of several research and scientific studies. It is believed that lectins play an important role in plant defense mechanisms, mainly due to their multivalence in recognizing and interacting with glycoconjugates from other organisms. Other characteristics that corroborate with this hypothesis are the fact that the production of plant lectins seems to be preferentially associated with regions susceptible to attack by predators, in addition to their high stability, even when subjected to unfavorable conditions of pH, temperature, and exposure to proteases from other organisms (Silva, 2008; Peumans and Van Damme, 1995).

According to Van Damme et al. (1998), some plant lectins (present in seeds and other reserve organs) have adaptive evolutionary characteristics that give them a dual function. These special proteins participate in nutrient storage processes, as well as defense, being toxic to predatory organisms.

Anti-insect Activity

Studies that evaluate the action of plant lectins against pathogenic insects demonstrated that these proteins represent promising insecticidal agents in insects belonging to different orders, including Lepidoptera, Coleoptera, and Hemiptera (Vandenborre, Smagghe and Van Damme, 2011). Lectins with mannose specificity and isolated from monocotyledonous plants are the most commonly reported with insecticidal properties, causing an increase in mortality or delaying insect development. The entomotoxic action of lectins is the result of their ability to interact with carbohydrates present in the digestive tract of insects (Lam and Ng, 2011; Silva, 2008). Bacterial heterologous systems are efficient for the expression of lectins with insecticidal potential. After purification and isolation, its biological activity can be evaluated through direct experiments – by introducing the proteins into the artificial diet of the target insects – or indirect – by topical application of a solution containing the lectin on the leaves of infected plants (Luo, Zhangsun and Tang, 2005). According to Macedo et al. (2015), the ingestion of lectins can cause antinutritional effects, which affect biological parameters in insects, such as larval weight, fecundity, pupation, and survival.

The possibility of incorporating exogenous genes and obtaining transgenic plants through genetic engineering have generated promising perspectives, regarding the possibility of transforming economically important plants and reducing the susceptibility of cultivars to pest attack (Sauvion et al., 2004).

Antifungal Activity

Despite the wide variety of plant lectins described, few have antifungal activity. Some studies indicate the fungal cell wall as the main target of lectins; however, the mechanism by which they exert their antifungal action is poorly known (Del Rio, Canal and Regente, 2020). Studies involving the exposure of vascular cells from roots and stems of *Gastrodia elata*, a monocotyledon belonging to the Orchidaceae family, to the fungus *Trichoderma viride* revealed an increase in the expression of lectins with specificity to mannose, indicating their possible relationship with the plant's defense system (Sa et al., 2003). In this case, transgenesis also represents a promising biotechnological tool for obtaining plants resistant to attack by phytopathotenic fungi.

According to Lam and Ng (2011), that lectins are not able to bind directly to glycoconjugates that are part of fungal membranes or to penetrate the cytoplasm due to the presence of the cell wall. Thus, it is unlikely that lectins act directly on the permeability or structure of the fungal membrane. However, indirect effects can be produced from the interaction of lectins with carbohydrates present on the surface of the cell wall inhibiting the development of these organisms.

Antitumor and Anticarcinogenic Activities

It is well documented that some plant lectins exhibit antitumor (inhibition of tumor growth) and anticarcinogenic (inhibition of cancer-induction by carcinogenic agents) activity. The ability of lectins to bind to carbohydrates with relative specificity allows interaction with glycoconjugates markedly expressed by certain tumor and cancer cells. The mechanisms of action of lectins against tumor cells are diverse, and the most common is the induction of apoptosis in various cell lines. Although the apoptotic pathways are different, activation of different caspases and cellular receptors usually occurs. Other mechanisms that have already observed are changes in cell proliferation and differentiation, production of pro-oxidant compounds, the release of

Cytochrome C, and depolarization of the mitochondrial membrane (Lawanprasert et al., 2020; Lam and Ng, 2011; Kim et al., 2001).

Lectins have been used for the early diagnosis of prostate cancer since some lectins are able to bind to glycoconjugates normally expressed in an altered way in tumor cells of this type (Lam and Ng, 2011). Gupta et al. (2022) reported the use of glycosylated plant lectins in targeted nanonovacins for cancer immunotherapy. According to the researchers, glycan domains of lectins – associated with antigens complexed with a nanocarrier of choice – are able to confuse immune cells and direct them to cross-presentation of antigens. This strategy is considered to be a better therapeutic alternative, when compared to monoclonal antibodies due to its cost-effectiveness, high specificity, toxicity, and reduced side effects.

Although studies have scientifically proven the potential of plant lectins in oncological therapies for use within the clinical setting, it is necessary to optimize protocols to reduce toxicity, improve efficiency, and effectiveness, in order to comply with regulations that guide good manufacturing practices for biopharmaceuticals (Papathanasiou and Kontoravdi, 2020; Bretas et al., 2010).

Antimicrobial Activity

Several studies describe the inhibitory effect of plant lectins on the growth of bacteria, fungi, and viruses. Although, the mechanisms of action employed by lectins are still poorly understood. According to Camacho (2007), the formation of ion channels in membranes and competitive inhibition by receptors may be involved.

In virology, the effects of lectins on the human immunodeficiency virus (HIV) represent one of the best investigated. Different plant lectins have different anti-HIV mechanisms, including inhibition of syncytia formation, inhibition of fusion of infected cells with CD4 defense cells, blocking virus entry into the cell, inhibition of reverse transcriptase enzyme, and viral replication (Lam and Ng, 2011; Swanson et al., 2010; Hansen et al., 1989). Recently, mannose-binding plant lectins have also exhibited anti coronavirus activity in cases of severe acute respiratory syndrome. These proteins intervene in the initial phase of the reproductive cycle, suppressing the viral multiplication cycle (Naik and Kumar, 2022; Capell et al., 2020; Lam and Ng, 2011; Keyaerts et al., 2007).

Furthermore, antibacterial and antinematoid activities have also been described for plant lectins. The antibacterial action of lectins results from the interaction with a range of carbohydrates present in the cell wall of some bacteria, such as teichoic acid, peptidoglycans, lipopolysaccharides, and muramic acid (Cavada et al., 2020; Zhang et al., 2020; Wang et al., 2005). In interactions of parasitism and infectious processes, the interaction between lectins and glycoconjugates stimulates the innate immune response in the host organism (Camacho, 2007).

Toxicity in Eukaryotic Cells

There is a group of highly toxic plant lectins capable of catalytically inactivating eukaryotic cell ribosomes. Its mechanism of action consists of the enzymatic removal of the adenine residue present in a highly conserved region of the larger ribosomal RNA subunit. Cleavage of this adenine residue causes conformational changes in the affected loop, preventing its interaction with elongation factors and consequent impairment of protein synthesis and cell death. These chimerolectins are called ribosome inactivating proteins (type II ribosome inactivating protein - RIP) and are formed by two subunits: the toxic A subunit, enzymatically active, and the B subunit with specificity to one or more carbohydrates, which acts as a lectin and serves as a bridge for the toxic subunit to enter the cell (Peumans and Van Damme, 1995).

Conclusion

Over the years, studies have demonstrated the complexity of plant lectins and their biological role. The discovery that many lectins are chimerical in nature raises the need to reassess other biological functions in which lectins may be involved. In addition, despite the broad biotechnological potential of plant lectins, the difficulty in obtaining isolated functional proteins still represents a challenge for carrying out more detailed histochemical, biochemical, and medicinal studies of their biological properties/applicability. Thus, the development of efficient systems and protocols for expression, detection, isolation, and purification of recombinant lectins represents an important step towards a better understanding of the true biochemical and physiological role of these proteins.

Disclaimer

None.

Acknowledgments

The authors are grateful to the Universidade Federal de São João del-Rei (UFSJ), Fundação de Amparo à Pesquisa do Estado de Minas Gerais (FAPEMIG - APQ-00861-22), Conselho Nacional de Desenvolvimento Científico e Tecnológico (CNPq – Processo 306581/2020-5) and Coordenação de Aperfeiçoamento de Pessoal de Nível Superior (CAPES) for financial support. This study was financed in part by the Coordenação de Aperfeiçoamento de Pessoal de Nível Superior – Brazil (CAPES) – Finance Code 001.

References

Azzoni, Adriano R. 2002. "*Plantas como biorreatores: recuperação e purificação de aprotinina recombinante a partir de semente de milho transgênico.*" [*Plants as bioreactors: recovery and purification of recombinant aprotinin from transgenic corn seed*] PhD diss., Campinas State University.

Bonnardel, Francois. 2021. "*Etude bioinformatique des lectines: nouvelle classification et prédiction dans les génomes.*" [*Bioinformatic study of lectins: new classification and prediction in genome*] PhD diss., Grenoble Alps University.

Boyd, William C. and Elizabeth Shapliegh. 1954. "Specific Precipitating Activity of Plant Agglutinins (Lectins)." *Science* 119:419. Accessed February 16, 2023. doi: 10.1126/science.119.3091.419.a.

Bretas, Rodrigo M., Bernardo L. M. Moreira, Brenda G. Valente, Daniela M. Cerqueira, Juliana B. da Silva, Marcelo M. M. Moreira, Marcos F. G. Silva, Maria F. R. S. Thees, Neemias S. Andrade, Raimundo N. V. Junior, Silmara C. Silveira and Laura G. Castanheira. 2010. *Proposta de guia para o estabelecimento e caracterização de bancos de células visando à produção de medicamentos biológicos.*" [*Proposed guide for the estabilishment and characterization of cells banks aimed at the production of biological medicines*] Accessed March 14, 2023. https://www.paho.org/hq/dm documents/2011/BRA--PROPOSTA-DE-GUIA-PARA-O-ESTABLECIMENTO-E-CARACTERIZACAO-DE-BANCO-DE-CELULAS.pdf.

Bucher, Matthew H., Artem G. Evdokimov and David S. Waugh. 2002. "Differential effects of short affinity tags on the crystallization of *Pyrococcus furiosus* maltodextrin-binding protein." *Acta Crystallographica D Biological Crystallography* 58(3):392–97. doi: 10.1107/s0907444901021187.

Camacho, Natália N. 2007. "*Expressão heteróloga da lectina de Bauhinia forficata Link em Escherichia coli e efeito sobre linhagens celulares tumorais*." [*Heterologous expression of Bauhinia forficata Link lectin in Escherichia coli and effect on tumor cell lines*]. Msc diss., Federal University of Pelotas.

Capell, Teresa, Richard M. Twyman, Victoria Armario-Najera, Julian K. C. Ma, Stefan Schillberg and Paul Christou. 2020. "Potential Applications of Plant Biotechnology against SARS-CoV-2." *Trends in Plant Science* 25(7):635–43. doi: 10.1016/j.tplants. 2020.04.009.

Cavada, Benildo S., Vanir R. Pinto-Junior, Vinicius J. S. Osterne, Messias V. Oliveira, Claudia F. Lossio, Mayara T. L. Silva, Alfa U. Bari, Lara D. Lima, Claudio H. D. Souza-Filho and Kyria S. Nascimento 2020. "Comprehensive review on Caelsalpinioideae lectins: From purification to biological activities." *International Journal of Biological Macromolecules* 162:333–48. Accessed February 15, 2023. doi: 10.1016/j.ijbiomac.2020.06.161.

Chant, Alan, Christina M. Kraemer-Pecorec, Richard Watkinb ana Geoff G. Knealea. 2004. "Attachment of a histidine tag to the minimal zinc finger protein of the *Aspergillus nidulans* gene regulatory protein AreA causes a conformational change at the DNA-binding site. *Protein Expression and Purification* 39(2):152–59. doi: 10.1016/ j.pep.2004.10.017.

Del Rio, Mariangela, Laura de la Canal and Mariana Regente. 2020. "Plant Antifungal Lectins: Mechanism of Action and Targets on Human Pathogenic Fungi." *Curr Protein Pept Sci.* 21(3):284-94. doi: 10.2174/1389203720666190906164448.

Goldstein, Irwin J., Harry C. Winter, Jennifer Aurandt, Laura Confer, Julie T. Adamson, Kristina Hakansson and Henriette Remmer. 2007. "A new α-galactosyl-binding protein from the mushroom Lyophyllum decastes." *Archives of Biochemistry and Biophysics* 467(2):268–74. doi: 10.1016/j.abb.2007.08.017.

Gupta, Bhavika, Daizy Sadaria, Vaishnavi U. Warrier, Anuradha Kirtonia, Ravi Kant, Amit Awasthi, Prakash Baligar, Jayanta K. Pal, Eiji Yuba, Gautam Sethi, Manoj Garg and Rajesh K. Gupta. 2022. "Plant lectins and their usage in preparing targeted nanovaccines for cancer immunotherapy." *Seminars in Cancer Biology* 80:87-106. doi: 10.1016/j.semcancer.2020.02.005.

Hajtó, Tibor, Katarina Hostanska, Timea Berki, László Pálinkás, Ferenc Boldizsár and Péter Németh. 2005. "Oncopharmacological perspectives of a plant lectin (Viscum album agglutinin-I): Overview of recent results from *in vitro* experiments and *in vivo* animal models, and their possible relevance for clinical applications." *Evidence-based Complementary and Alternative Medicine* 2(1):59–67. Accessed February 20, 2023. doi: 10.1093/ecam/neh058.

Hansen, John E., C. M. Nielsen, C. Nielsen, P. Heegaard, L. R. Mathiesen and J. O. Nielsen. 1989. "Correlation between carbohydrate structures on the envelope glycoprotein gp120 of HIV-1 and HIV-2 and syncytium inhibitionwith lectins." *Aids* 3(10):635-41. doi: 10.1097/00002030-198910000-00003.

Islam, Md R., Seoyoung Choi, Thangarasu Muthamilselvan, Kunyoo Shin and Inhwan Hwang. 2020. "*In Vivo* Removal of N-Terminal Fusion Domains from Recombinant

Target Proteins Produced in *Nicotiana benthamiana*." *Frontiers in Plant Science* 11:1–14. Accessed March 3, 2023. doi: 10.3389/fpls.2020.00440.

Jung, Eui C., Ki D. Kim, Chan H. Bae, Ju C. Kim, Dae K. Kim and Ha H. Kim. 2007. "A mushroom lectin from ascomycete *Cordyceps militaris*." *Biochimica et Biophysica Acta* 1770(5):833–38. doi: 10.1016/j.bbagen.2007.01.005.

Keyaerts, Els, Leen Vijgen, Christophe Pannecouque, Els Van Damme, Willy Peumans, Herman Egberink, Jan Balzarini and Marc Van Ranst. 2007. "Plant lectins are potent inhibitors of coronaviruses by interfering with two targets in the viral replication cycle." *Antiviral Research* 75(3):179–87. Accessed March 15, 2023. doi: 10.1016/j.antiviral.2007.03.003.

Kim, M. Sunny, Jienny Lee, Hong-Seob So, Kang-Min Lee, Byung-Hak Jung, Sang-Young Chung, Sun-Rock Moon, Nam-Song Kim, Chang-Bo Ko, Hye-Jung Kim, Yong-Kyu Kim and Raekil Park. 2001. "Gamma-interferon (IFN-γ) augments apoptotic response to mistletoe lectin-II via upregulation of Fas/Fas L expression and caspase activation in human myeloid U937 cells." *Immunopharmacology and Immunotoxicology* 23:55–66. doi: 10.1081/iph-100102567.

Lam, Sze K., Ng, Tzi B. 2011. "Lectins: Production and practical applications." *Applied Microbiology and Biotechnology* 89(1):45–55. doi: 10.1007/s00253-010-2892-9.

Lannoo, Nausicaä, Wouter Vervecken, Paul Proost, Pierre Rougé and Els J. M. Van Damme. 2007. "Expression of the nucleocytoplasmic tobacco lectin in the yeast *Pichia pastoris*." *Protein Expression and Purification* 53(2):275–82. doi: 10.1016/j.pep.2007.01.007.

Lawanprasert, Atip, Caitlin A. Guinan, Erica A. Langford, Carly E. Hawkins, Janna N. Sloand, Howard W. Fescemyer, Matthew R. Aronson, Jacob A. Halle, James H. Marden, Scott H. Medina. 2020. "Discovery of antitumor lectins from rainforest tree root transcriptomes." *Plos One* 15(2): 1–17. Accessed March 5, 2023. doi: 10.1371/journal.pone.0229467.

Luo, Sulan, Zhangsun, Dongting and Tang, Kexuan. 2005. "Functional GNA expressed in *Escherichia coli* with high efficiency and its effect on *Ceratovacuna lanigera* zehntner." *Applied Microbiology and Biotechnology* 69(2):184–91. Accessed March 3, 2023. doi: 10.1007/s00253-005-0042-6.

Macedo, Maria L. R., Caio F. R. Oliveira and Carolina T. Oliveira. 2015. "Insecticidal activity of plant lectins and potential application in crop protection." *Molecules* 20:2014–33. Accessed March 16, 2023. doi: 10.3390/molecules20022014.

Naik, Sanjay and Sanjit Kumar. 2022. "Lectins from plants and algae act as anti-viral against HIV, influenza and coronaviruses." *Molecular Biology Reports* 49:12239–46. Accessed March 16, 2023. doi: 10.1007/s11033-022-07854-8.

Oliveira, Carla, Wagner Felix, Renato A. Moreira, José A. Teixeira and Lucília Domingues. 2008. "Expression of frutalin, an α-d-galactose-binding jacalin-related lectin, in the yeast *Pichia pastoris*." *Protein Expression and Purification* 60(2):188–93. doi: 10.1016/j.pep.2008.04.008.

Papathanasiou, Maria M. and Kontoravdi, Cleo. 2020. "Engineering challenges in therapeutic protein product and process design." *Current Opinion in Chemical Engineering* 27:81–88. Accessed March 15, 2023. doi: 10.1016/j.coche.2019.11.010.

Peumans, Willy J. and Els J. M. Van Damme. 1995. "Lectins as plant defense proteins." *Plant Physiology* 109(2):347–52. Accessed February 21, 2023. doi: 10.1104/pp.109.2.347.

Pietrzyk-Brzezinska, Agnieszka J. and Anna Bujacz. 2020. "H-type lectins – Structural characteristics and their applications in diagnostics, analytics and drug delivery." *International Journal of Biological Macromolecules* 152:735–47. Accessed March 2, 2023. doi: 10.1016/j.ijbiomac.2020.02.320.

Puhl, Ana C. 2008. "*Obtenção de um biocatalisador da protease TEV para a remoção de caudas de histidinas de proteínas recombinantes*." [*Obtaining a TEV protease biocatalisty for the removal of hitydine tags from recombinant proteins*]. PhD diss., Federal University of Santa Catarina.

Radha, P and Rajasekhar Urla. 2020. "Cloning and characterization of mannose binding jacalin related lectin encoding gene from Indian *Cycas* (*Cycas annaikalensis* Singh and Radha)." *Journal of Plant Biochemistry and Biotechnology* 29(2):314–22. Accessed March 10, 2023. doi: 10.1271/bbb.60156.

Raemaekers, Romaan J. M., Laura de Muro, John A. Gatehouse and Anthony P. Fordham-Skelton. 1999. "Functional phytohemagglutinin (PHA) and *Galanthus nivalis* agglutinin (GNA) expressed in *Pichia pastoris*. Correct N-terminal processing and secretion of heterologous proteins expressed using the PHA-E signal peptide." *European Journal of Biochemistry* 265:394–403. Accessed March 2, 2023. doi: 10.1046/j.1432-1327.1999.00749.x.

Sa, Q., Y. Wang, W. Li, L. Zhang and Y. Sun. 2003. The promoter of an antifungal protein gene from *Gastrodia elata* confers tissue-specific and fungus-inducible expression patterns and responds to both salicylic acid and jasmonic acid." *Plant Cell Reports* 22:79–84. doi: 10.1007/s00299-003-0664-z.

Sauvion, Nicolas, Hubert Charles, Gérard Febvay and Yvan Rahbé. 2004. "Effects of jackbean lectin (ConA) on the feeding behaviour and kinetics of intoxication of the pea aphid, *Acyrthosiphon pisum*." *Entomologia Experimentalis et Applicata* 110:31–44. Accessed March 14, 2023. doi: 10.1111/j.0013-8703.2004.00117.x.

Sharon, Nathan and Halina Lis. 2004. "History of lectins: from hemagglutinins to biological recognition molecules." *Glycobiology* 14, 11:53–62. Accessed February 15, 2023. doi: 10.1093/glycob/cwh122.

Silva, Michelle D. C. 2008. "Aplicações Biotecnológicas das Lectinas ClaveLL (*Cladonia verticillaris* Lichen Lectin) e BmoLL (*Bauhinia monandra* Leaf Lectin)." [Biotechnological Applications of ClaveLL (*Cladonia verticillaris* Lichen Lectin) and BmoLL (*Bauhinia monandra* Leaf Lectin)]. PhD diss., Federal University of Pernambuco.

Silva, Felipe F., Daniel B. Gonçalves and Lopes, Débora O. "The use of bioinformatics tools to characterize a hypothetical protein from *Penicillium rubens*." *Genetics and Molecular Research* 19 (2):1–18. Accessed March 03, 2023. doi: //dx.doi.org/10.4238/gmr18574.

Swanson, Michael D., Harry C. Winter, Irwin J. Goldstein and David M. Markovitz. 2010. "A lectin isolated from bananas is a potent inhibitor of HIV replication." *Journal of Biological Chemistry* 285(12):8646–55. Accessed March 15, 2023. doi: 10.1074/jbc.M109.034926.

Tsaneva, Mariya; Els J. M. Van Damme. 2020. "130 years of Plant Lectin Research." *Glycoconjugate Journal* 37:533–51. Accessed March 16, 2023. doi: 10.1007/s10719-020-09942-y.

Tyurin, Alexander A., Alexandra V. Suhorukova, Ksenia V. Kabardaeva and Irina V. Goldenkova-Pavlova. 2020. "Transient gene expression is an effective experimental tool for the research into the fine mechanisms of plant gene function: Advantages, limitations, and solutions." *Plants* 9(9):1–19 Accessed March 5, 2023. doi: 10.3390/plants9091187.

Van Damme, Els J. M., Willy J. Peumans, Annick Barre and Pierre Rougé. 1998. "Plant lectins: A composite of several distinct families of structurally and evolutionary related proteins with diverse biological roles." *Critical Reviews in Plant Sciences* 17(6) 575–692. Accessed February 20, 2023. doi: 10.1080/07352689891304276.

Van Damme, Els J. M. 2014. "History of Plant Lectin Research". In *Lectins: Methods and Protocols, Methods in Molecular Biology*, vol.1200, edited by Jun Hirabayashi, 3-13. New York: Humana Press.

Vandenborre, G., Guy Smagghe and Van Damme, Els J. M. 2011. "Plant lectins as defense proteins against phytophagous insects." *Phytochemistry* 72:1538–50. doi: 10.1016/j.phytochem.2011.02.024.

Wang, Zhaoyu, Kewei Zhang, Xiaofen Sun, Kexuan Tang and Juren Zhang. 2005. "Enhancement of resistance to aphids by introducing the snowdrop lectin gene gna into maize plants." *J. Biosci.* 30:627–38. Accessed March 15, 2023. doi: 10.1007/BF02703563.

Zhang, Changwei, Huiyu Wang, Yuanyuan Xu, Shuning Zhang, Jianjun Wang, Bing Hu, Xilin Hou, Ying Li and Tongkun Liu. 2020. "Enhanced Relative Electron Transport Rate Contributes to Increased Photosynthetic Capacity in Autotetraploid Pak Choi." *Plant and Cell Physiology* 61(4):761–74. doi: 10.1093/pcp/pcz238.

Chapter 5

The Use of Multivariate Statistical Methods in the Analysis of Histoenzymatic Activity of the Skin Glands in Rats

**A. B. Kiladze*
and N. K. Dzhemukhadze**
Laboratory of Ecology, Physiology and Functional Morphology of Higher Vertebrates, A. N. Severtsov Institute of Ecology and Evolution, Russian Academy of Sciences, Moscow, Russia

Abstract

The translation of the sign form of histoenzymatic activity into the form of digital analogues allowed the use of methods of multivariate statistics in order to identify significant differences between the histochemical activity of enzymes of the skin glands of laboratory and wild rats. It was found that the activity of acid phosphatase and alkaline phosphatase was slightly higher compared to the activity of adenosine triphosphatase, while the activity of peroxidase differed by minimal values. Discriminant analysis revealed clear differences between White rats, August and Wistar rats, as well as Norway rats, which is demonstrated in the scattering diagram. The parameters of classification equations for laboratory and wild rats are determined. It is shown that the created classification of rats based on the histoenzymatic parameters of the skin glands is generally correct by 75%, while the most correct level of data discrimination is characteristic of Norway rats (100% correct attribution), and the lowest is for August rats (only 40% correct attribution), 80% of the correct data are White rats and Wistar rats.

---

* Corresponding Author's Email: andreykiladze@yandex.ru.

In: Advances in Biology. Volume 4
Editor: Charles D. Grant
ISBN: 979-8-88697-999-2

Cluster analysis revealed the maximum level of similarity of histochemical activity of enzymes in the skin glands of August and Wistar rats, while the maximum distance is observed between Wistar rats and Norway rats.

Keywords: laboratory and wild rats, histochemistry of enzymes, skin glands, graphical analysis of variability, discriminant analysis, cluster analysis

Introduction

Modern histoenzymology as a separate scientific discipline (Meier-Ruge and Bruder, 2008; Wick, 2012) requires the use of various methods of processing the result, which can be both directly quantitative in nature, providing an analytical component to the study, and semi-quantitative in nature, for example, when staging histochemical reactions, the results of which, under microscopic observation, can be interpreted in the following ways: (*i*) the presence or absence of an enzyme, which is a reflection of alternative signs; (*ii*) evaluate the enzyme content in the verbal formulations "low" or "high" (Ageev, 1969), and (*iii*) identify gradations of activity, which requires a high level of qualification and experience in histochemistry (Nachlas et al., 1958; Sokolov et al., 1988). It is the latter form of visual observation of histoenzymatic activity that can be considered the most progressive method of semi-quantitative analysis (Dzhemukhadze, 2007a).

As a rule, in light histoenzymology, the gradations of enzyme activity are traditionally expressed in a sign form ("+" and/or "–"), while in our long-term practice, the scale of histoenzymatic activity has the following form: "–", "–(+)", "+(–)", "+", "++", "+++", that is, from the complete absence of the enzyme or the presence of its traces ("–") to high enzyme activity ("+++") (Dzhemukhadze, 2007a).

At the same time, the most rational way to represent a semi-quantitative gradation of the histochemical result is a digital form, since it becomes possible to use statistical methods (Dzhemukhadze and Kiladze, 2008; Kiladze and Dzhemukhadze, 2013). The translation of the sign form of activity into the form of digital analogues in the form of points based on qualimetry methods (Azgaldov and Kostin, 2011; Lobanov, 2013; Rozhkov, 2018) allowed us to introduce an appropriate ordinal scale earlier (Stevens, 1946; Marusina, Tkalich and Labkovskaya, 2019), combining a verbal and visual representation of enzyme activity, a sign expression and a corresponding digital expression of activity (Figure 1).

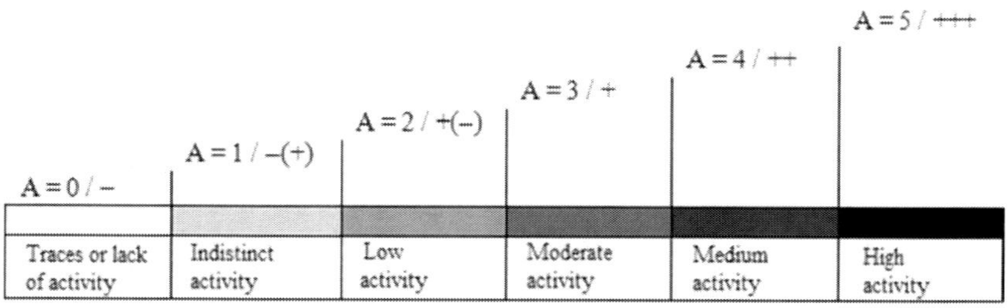

Figure 1. The ordinal scale of histoenzymatic activity in digital form. *Note*: A — enzyme activity in the form of digital analogues (points) and signs (Kiladze and Dzhemukhadze, 2022a).

The use of digital analogues to evaluate histoenzymatic activity is a quantification of data (Wilks, 1961), which allows for quantitative expression of qualitative characteristics. It should be noted that the representation of histochemical activity of enzymes in the form of digital analogues (in points) is increasingly used in the practice of histochemistry and immunohistochemistry (Warburton and Santer, 1994; Amirova, 2018; Bencze et al., 2022).

Considering that histoenzymatic preparations are not able to persist for a long time, since enzymes are rapidly destroyed, from the point of view of forming a reliable histochemical archive, it is extremely important to fix the histochemical result in a short time, presenting it in the form of gradations of enzyme activity (Sokolov and Dzhemukhadze, 1982).

Previously, we have already used a wide variety of statistical, biomathematics and bioinformatics approaches, which became possible through the use of digital analogues in enzyme histochemistry, while solving specific biological problems in the field of chemocommunication (Dzhemukhadze and Kiladze, 2012; Kiladze and Dzhemukhadze, 2016) and interspecific interactions (Kiladze and Dzhemukhadze, 2020, 2021a), evaluation of sexual dimorphism (Kiladze and Dzhemukhadze, 2019, 2021b,c) and seasonal variability (Kiladze and Dzhemukhadze, 2022b), as well as in taxonomic studies (Dzhemukhadze and Kiladze, 2014).

Thus, this study is a logical continuation of earlier work related to the use of statistical methods in histochemical practice. The purpose of this work is to test discriminant analysis (Fisher, 1936) related to the group of methods of multivariate statistics (Marcoulides and Hershberger, 1997).

Discriminant analysis is supposed to be subjected to histochemical activity of the skin glands of laboratory and Norway rats, the first of which are widely used in experimental studies as model animals in biomedical work (Hashimoto and Ogawa, 1963; Niczyporuk, 2018). Comparison of laboratory rats with Norway rats belonging to wild rodents is intended to show differences in the metabolism of these groups of animals. It should be noted that laboratory rats are the domesticated descendants of the wild form of the Norway rat *Rattus norvegicus* Berkenhout, 1769 (Sharp and Villano, 2012; Pritchett-Corning, 2015). We used outbred White rats, as well as two inbred forms — August rats and Wistar rats (Hedrich, 2000).

In addition, Norway rats belonging to synanthropic species (Mazzotta et al., 2023), cosmopolitan species (Uea-Anuwong et al., 2023) and invasive species (King, Foster and Miller, 2011; Canale et al., 2019) are distinguished by worldwide settlement, therefore it is necessary to restrain the number of wild rats, since they cause significant harm to human economic activity by their vital activity, as well as spread infectious diseases, presenting an epidemiological danger (Soh et al., 2023).

In this regard, knowledge of the histochemical activity of the skin glands involved in marking behavior, which forms a favorable background for successful reproduction (Desvars-Larrive et al., 2018; Byers et al., 2019), is an important biological aspect in this activity, since pest management is based on biological patterns of management of populations of synanthropic rodents (Klemann et al., 2023), which has become especially relevant during the active phase of the pandemic of the novel coronavirus infection (COVID-19). Thus, the introduction of nationwide lockdowns led to a shift in the behavioral pattern of Norway rats. These rats living in the largest megacities of the world (Harbison, 2020) have become more aggressive and active even in the daytime, which was due to a significant reduction in the food supply, which has an anthropogenic character (food waste), since catering establishments were closed everywhere (Parsons et al., 2020; Searle, Turnbull and Lorimer, 2021).

It is obvious that the above facts, both from a scientific and practical point of view, make it relevant to conduct biological studies on the histoenzymology of laboratory and wild rats.

Thus, this work is intended to reflect the histoenzymatic differences between the skin glands of laboratory and wild rats using discriminant analysis methods, which can be considered the first such attempt to use this kind of multivariate statistical approach in our research.

Material and Methods

Laboratory and wild Norway rats in the following number were selected as the object of the study: (1) outbred White rats *Rattus norvegicus* var. *alba* — 30 individuals; (2) inbred August rats *Rattus norvegicus* Aug / Lac Sto — 20 individuals; (3) inbred Wistar rats *Rattus norvegicus* Wag / G Sto — 20 individuals; (4) Norway rats *Rattus norvegicus* Berkenhout, 1769 — 7 individuals (Dzhemukhadze, 2007a). Only mature adult males of the same season were selected, which can be explained by their dominant position in relation to females, which is especially characteristic of Norway rats (Dzhemukhadze, 2007b). Laboratory rats were provided by the vivarium of the Severtsov Institute of Ecology and Evolution of the Russian Academy of Sciences, while Norway rats were caught in Moscow. The sebaceous glands of the skin were examined from the following topographic sites: nape, mouth corners, eyelids, anal area, as well as the eccrine glands of the soles of the paws (Sokolov and Chernova, 2001).

Skin samples were fixed in 10% neutral formalin in the cold for 8-10 hours. The thickness of the frozen "floating" sections was 15 μm. The histochemical study included the formulation of reactions to acid phosphatase (ACP), alkaline phosphatase (ALP), adenosine triphosphatase (ATPase), as well as peroxidase (Dixon and Webb, 1982). The reaction to ACP was carried out by the Gomori method, on the ALP — by the method with metal salts according to Burstone. The ATPase was determined using a modification of the classical lead Gomori method with magnesium ions in an incubation medium. Peroxidase was detected using the M-Nadi reaction (Gomori, 1952; Pearse, 1960; Burstone, 1962).

For better visualization and the possibility of detecting gradations of histoenzymatic activity, freshly prepared preparations were not stained. According to the preparations, the degree of enzyme activity was assessed by visual signs (sediment density, as well as the area occupied by it in the field of view of the microscope). The histochemical activity of the studied enzymes manifests itself mainly in the form of grains, granules and diffusely colored areas in the cytoplasm of the cell, having different intensity of manifestation on the preparation (Montagna and Noback, 1947; Lojda, Gossrau and Schiebler, 1979; Luppa, 1980; Sokolov and Dzhemukhadze, 1982).

Histochemical activity of enzymes was expressed as digital analogues (in points) using an ordinal scale shown in Figure 1.

To reflect the variability of enzyme activity in the studied rodents, a rectangular coordinate system was used, where laboratory and wild rats were indicated along the abscissa axis, and digital analogues of histoenzymatic activity were indicated along the ordinate axis, indicating activity variants and the limits of variability.

For discriminant analysis (McLachlan, 2004), a matrix was compiled that included 80 values of digital analogues of histoenzymatic activity, which was formed by 4 types of enzymes, 4 forms of rats, including 5 cases each, which is equivalent to 5 topographic areas. Thus, the matrix was formed by 20 cases. All four variables reflecting the activity of ACP, ALP, ATPase and peroxidase were included in the model. Digital analogues of histochemical activity of enzymes are independent variables, while the grouping variable is 4 forms of rats (White rats, August rats, Wistar rats and Norway rats). To confirm the results of the discriminant analysis, a cluster analysis was performed using the Ward's method (Gower, 1967; Everitt, Landau and Leese, 2001). For the purposes of cluster analysis, the data matrix has been rebuilt. The statistical analysis was carried out on the basis of the computer program "STATISTICA 10" (StatSoft, USA) (Khalafyan, 2008).

Results

At the first stage, we will translate the sign form of the enzyme activity of the skin glands of laboratory and wild rats into digital analogues, as shown in Table 1.

The obtained digital analogues of enzyme activity make it possible to identify various biological patterns using a wide range of statistical methods.

At the second stage, we will consider the variability of the activity of enzymes of the skin glands of rats in a comparative aspect (Figure 2).

It was shown that identical limits of variability and variants of ACP activity (from 3 to 5 points) were revealed in White rats and August rats. Wistar rats according to ACP have no variability, since the activity of enzymes of the skin glands in all topographic areas was equally high (5 points). The activity of ACP in Norway rats varied from 2 to 4 points, that is, the activity of this enzyme in wild rats is reduced compared to laboratory rats.

ALP in White rats had two variants of activity — 3 and 5 points, in the August rat, ALP varied with a smaller scale — 4 and 5 points. The variability of the activity of ALP in Wistar rats and Norway rats is identical — from 3 to 5 points with three variants of activity.

Table 1. Histoenzymatic activity on the example of the skin glands of male laboratory and wild rats in sign form (Dzhemukhadze, 2007a) and in the form of digital analogues (in points)

Enzymes	Topographic areas	White rats	August rats	Wistar rats	Norway rats
Acid phosphatase activity in the form of signs / points	Nape	+++ /5	++ / 4	+++ / 5	+ / 3
	Mouth corners				
	Upper eyelids	+ / 3	+++ / 5	+++ / 5	+ / 3
	Anal area	++ / 4	+++ / 5	+++ / 5	++ / 4
	The soles of the paws	+ / 3	+++ / 5	+++ / 5	+ / 3
		++ / 4	+ / 3	+++ / 5	+(−) / 2
Alkaline phosphatase activity in the form of signs / points	Nape	+ / 3	++ / 4	+ / 3	+ / 3
	Mouth corners				
	Upper eyelids	+++ /5	++ / 4	++ / 4	++ / 4
	Anal area	+ / 3	++ / 4	+++ / 5	+ / 3
	The soles of the paws	+++ /5	+++ / 5	++ / 4	+++ / 5
		+ / 3	+++ / 5	++ / 4	+++ / 5
Adenosine triphosphatase activity in the form of signs / points	Nape	+ / 3	+ / 3	+ / 3	−(+) / 1
	Mouth corners				
	Upper eyelids	++ / 4	+ / 3	+ / 3	+ / 3
	Anal area	++ / 4	+ / 3	+ / 3	+ / 3
	The soles of the paws	++ / 4	+ / 3	+ / 3	+ / 3
		+ / 3	+ / 3	+ / 3	++ / 4
Peroxidase activity in the form of signs / points	Nape	− / 0	− / 0	− / 0	− / 0
	Mouth corners				
	Upper eyelids	− / 0	− / 0	− / 0	+(−) / 2
	Anal area	+ / 3	+ / 3	+ / 3	+(−) / 2
	The soles of the paws	− / 0	− / 0	− / 0	− / 0
		− / 0	− / 0	− / 0	+(−) / 2

ATPase in White rats varied from 3 to 4 points, while in August and Wistar rats this enzyme was characterized by a stable moderate index (3 points) and no variability. In Norway rats, the range of ATPase activity is more significant — from 1 to 4 points with three variants of activity.

The histochemical activity of peroxidase in comparison with the three types of phosphatases in the skin glands of rats is significantly reduced. The dominant variant of activity was 0 points, which indicates the absence of this enzyme in the tissue and cellular structures of the skin glands of rats. In isolated cases, moderate activity (3 points) was recorded in laboratory rats and low activity (2 points) in Norway rats.

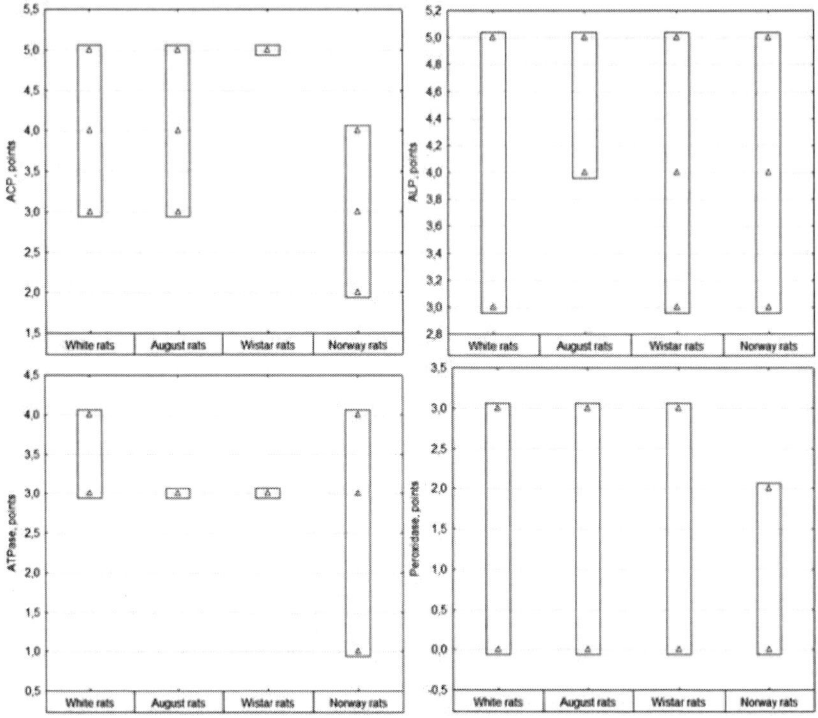

Figure 2. Variability of digital analogues of histoenzymatic activity of skin glands of rats. *Note*: The initial variants are shown by triangles, while the boundaries of variability are enclosed in rectangles.

In general, the histoenzymatic activity of ACP and ALP is higher than that of ATPase, while the manifestation of peroxidase activity is either absent or has moderate or low values in the skin glands of laboratory and wild rats.

At the third stage, we will consider the results of discriminant analysis, which is based on differences in the histoenzymatic activity of the skin glands of the studied rodents. It should be noted that the results obtained can have both a visual form in the form of a scattering diagram (Figure 3) and an analytical form characterizing the discrimination of data, their ratio relative to each other, as well as the level of correctness of the resulting classification.

For the primary evaluation of data discrimination, let's consider a scattering diagram, from which the distinction between outbred White rats, inbred rats (August and Wistar), as well as wild Norway rats is clearly visible (Figure 3).

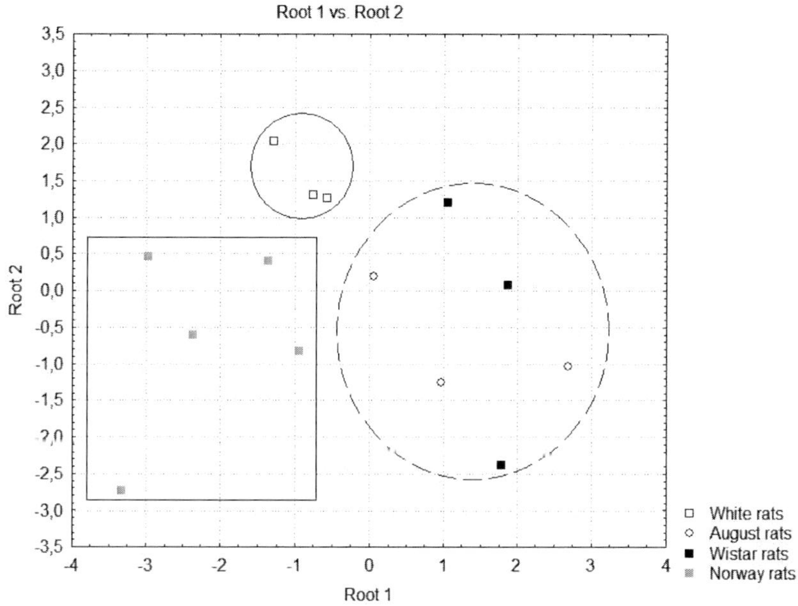

Figure 3. Results of discrimination of laboratory and wild rats based on histoenzymatic activity of the skin glands.

In this regard, we can confidently speak about the significance of histoenzymatic parameters reflecting the physiological differences between laboratory and wild rats and the possibility of using these parameters are used in various comparisons to solve various biological problems.

Consider the key parameters of discriminant analysis: (1) the number of variables in the model is 4; (2) Wilks' lambda is $\lambda = 0.1390$, which indicates good data discrimination, since the parameter value is approaching zero; (3) the F value for certain values of the number of degrees of freedom is approx. $F(12,34) = 3.2027$; (4) the significance level is $p < 0.0035$, which indicates the reliability of the data.

Classification equations for laboratory and wild rats can be represented as follows:

1. White rats = 19.83ACP + 10.19ALP + 12.08ATPase – 3.22Peroxidase – 79.21;
2. August rats = 22.46ACP + 13.68ALP + 8.53ATPase – 2.79Peroxidase – 92.88;

3. Wistar rats = 23.88ACP + 13.49ALP + 9.46ATPase – 3.07Peroxidase – 101.35;
4. Norway rats = 16.87ACP + 10.96ALP + 7.08ATPase – 1.73Peroxidase – 57.49.

It should be noted that the classification matrix reflecting the ratio between the 20 cases included in the discriminant analysis is quite accurate, as evidenced by the proportions of correct cases in Table 2.

Table 2. The level of correctness of classification of laboratory and wild rats by histoenzymatic activity of the skin glands

Laboratory and wild rats	The proportion of correct cases in the classification, %	White rats	August rats	Wistar rats	Norway rats
White rats	80.0	4	0	1	0
August rats	40.0	0	2	2	1
Wistar rats	80.0	0	1	4	0
Norway rats	100.0	0	0	0	5
Total	75.0	4	3	7	6

Speaking about the results obtained, it is necessary to point out only rather heterogeneous data related to August rats, which caused a low proportion of correct cases in the above classification, since only two cases out of five, according to discriminant analysis, belong to this group of inbred rats. The remaining cases can be considered more or less reliable in terms of data discrimination.

In this regard, the total level of attribution correctness (TLAC) for this classification is equal to TLAC = (20 – 5 / 20)100 = 75%.

In order to further investigate the relationship between laboratory and wild rats, a cluster analysis was performed using the Ward's method. It can be noted that the data of discriminant and cluster analyses are consistent, since the resulting dendrogram revealed similar separation trends between laboratory and wild rats, which is shown in Figure 4. It is noteworthy that inbred rats (August and Wistar) form a separate cluster, demonstrating the maximum similarity of histoenzymatic activity of the skin glands among themselves.

As analytical data of cluster analysis, we present the Euclidean distances between laboratory and wild rats, which allows us to understand the level of similarity and differences between the studied groups of rats (Table 3).

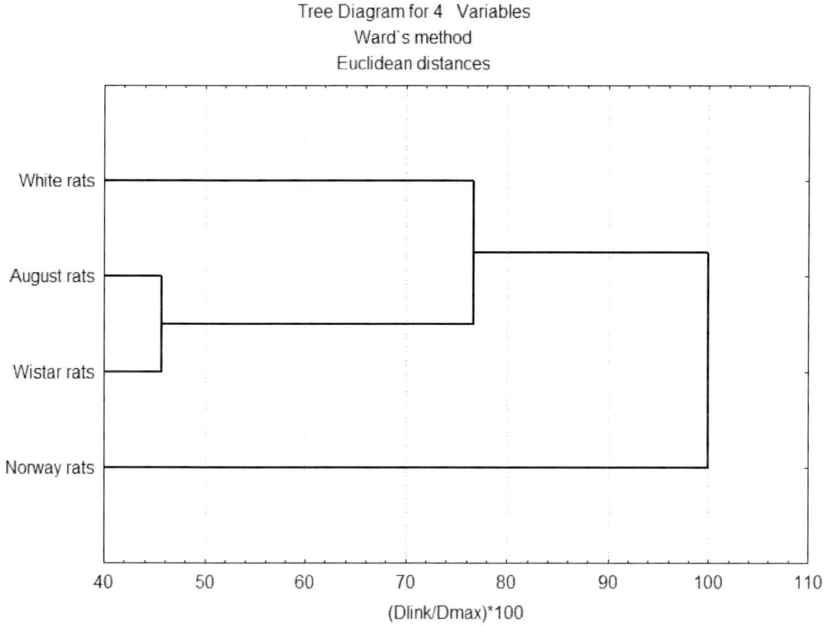

Figure 4. Results of clustering of laboratory and wild rats by the level of histoenzymatic activity of the skin glands.

Table 3. Euclidean distances between laboratory and wild rats by the level of histoenzymatic activity of the skin glands

Laboratory and wild rats	White rats	August rats	Wistar rats	Norway rats
White rats	0.00	4.58	4.47	5.48
August rats	4.58	0.00	3.00	5.20
Wistar rats	4.47	3.00	0.00	6.48
Norway rats	5.48	5.20	6.48	0.00

Thus, the Euclidean distances are fully consistent with the dendrogram reflecting the visual relationship between laboratory and wild rats.

Discussion

The choice of laboratory and wild rats as objects of research can be considered quite relevant, which is associated with the identification of differences in the intracellular metabolism of these rodents, which is due to the unequal histoenzymatic activity of the skin glands responsible for the formation of secretions (Saga and Morimoto, 1995; Saga, 2001, 2002; Cui and Schlessinger, 2015). In addition, differences in the histochemical activity of enzymes form an unequal physiological background of the glands of the skin involved in marking behavior, which is especially significant for wild Norway rats during the breeding season (Dzhemukhadze and Kiladze, 2013). Chemocommunication between individuals is carried out through various biological channels, one of which is the glandular apparatus of the skin, which acts as a source of chemical signals in animals (Sokolov and Chernova, 2001; Schweinfurth, 2020). Morphologically, the skin glands have a typical modular and repetitive structure (Chernova and Kiladze, 2019): the sebaceous glands have a cluster-like shape adjacent to the hair, while the eccrine sweat glands have the shape of a glomerulus with a tubular duct opening on the surface of the skin (Sokolov, 1982; Sokolov and Chernova, 2001).

The importance of laboratory rats is determined by their widespread use as model animals in laboratory diagnostics and in experimental studies (Dorsett-Martin, 2004; Hrytsevych and Vereschaka, 2021). Norway rats, on the contrary, as a synanthropic species, require constant control over the population size, since wild rats harm humans both in economic activities and in terms of the transmission of dangerous infectious diseases (Feng and Himsworth, 2014; Haniza et al., 2015; Barnett, 2017).

Turning to the discussion of the methods used to process the histochemical result, we once again draw attention to the fact that the translation of the sign form of activity into the form of digital analogues (Kiladze and Dzhemukhadze, 2013) allowed us to use various methods of multivariate statistics (Morrison, 2004). Thus, the graphical representation of variability visually showed the scope and limits of variability of histoenzymatic activity for four studied enzymes in a comparative series of laboratory and wild rats. Discriminant analysis made it possible to use histoenzymatic activity as classification parameters, on the basis of which there was a separation of data, which can be considered sufficiently justified from a biological point of view. It is shown that the level of correct classification for Norway rats is 100%, which indicates significant differences in the histoenzymatic activity of the skin glands of wild and laboratory rats. It should be noted that cluster analysis confirmed the identified trends found in the framework of discriminant analysis.

Conclusion

Thus, using digital analogues of histoenzymatic activity, it is possible to carry out the necessary statistical analysis of the data. Different levels of enzyme activity of the skin glands were used as the basis for discriminant analysis, which made it possible to correctly classify laboratory and wild rats by the nature of intracellular metabolism in the studied morphological structures. The resulting conclusions can be formulated in the form of the following theses:

1. different levels of enzyme activity with different limits of variability were determined, namely: the activity of ACP and ALP is higher than that of ATPase, while the manifestation of peroxidase activity was minimal;
2. discriminant analysis showed a clear separation between White rats, August and Wistar rats and Norway rats, which suggests the importance of histoenzymatic characteristics as classification features;
3. the use of cluster analysis showed that the minimum distance is observed between the histoenzymatic activity of the skin glands of inbred August and Wistar rats, and the maximum is between Norway rats and Wistar rats.

Disclaimer

None

References

Ageev A K. *Histochemistry of human alkaline and acid phosphatases in norm and pathology*. Leningrad: Meditsina; 1969. (In Russian)

Aminova G G. Acid phosphatase activity and the structure of neurons in the stellate ganglion in cattle. *Neuroscience and Behavioral Physiology* (2018) 48(5):571-576.

Azgaldov G, Kostin A. Applied qualimetry: its origins, errors and misconceptions. *Benchmarking: An International Journal* (2011) 18(3):428-444.

Barnett S A. *The rat: A study in behavior*. New York, NY: Routledge; 2017.

Bencze J, Szarka M, Kóti B, Seo W, Hortobágyi T G, Bencs V, Módis L V, Hortobágyi T. Comparison of semi-quantitative scoring and artificial intelligence aided digital image analysis of chromogenic immunohistochemistry. *Biomolecules* (2022), 12:19.

Burstone M S. *Enzyme Histochemistry and Its Application in the Study of Neoplasms.* New York, NY: Academic Press; 1962.

Byers K A, Lee M J, Patrick D M, Himsworth C G. Rats about town: A systematic review of rat movement in urban ecosystems. *Frontiers in Ecology and Evolution* (2019) 7:13.

Canale D E, Di Dio V, Massa B, Mori E. First successful eradication of invasive Norway rats Rattus norvegicus from a small Mediterranean island (Isola delle Femmine, Italy). *Folia Zoologica (*2019) 68(1):29-34.

Chernova O F, Kiladze A B. Heterochrony as the basis for inter- and intraspecific diversity of skin in vertebrates. *Biology Bulletin Reviews* (2019) 9(2):174–189.

Cui C-Y, Schlessinger D. Eccrine sweat gland development and sweat secretion. *Experimental Dermatology* (2015) 24(9):644-650.

Desvars-Larrive A, Baldi M, Walter T, Zink R, Walzer C. Brown rats (*Rattus norvegicus*) in urban ecosystems: are the constraints related to fieldwork a limit to their study? *Urban Ecosystems* (2018) 21:951-964.

Dixon M, Webb E. *Enzymes.* Moscow: Mir; 1982. (In Russian)

Dorsett-Martin WA. Rat models of skin wound healing: A review. *Wound Repair and Regeneration* (2004) 12(6):591-599.

Dzhemukhadze N K. Comparative phosphatase histochemistry in skin glands of laboratory and wild Norway rats (*Rattus norvegicus*). *Doklady Biological Sciences.* (2007a) 412(1):32-34.

Dzhemukhadze N K. The dependence of interspecific differences in the histoenzymatic parameters of skin glands between Norway (*Rattus norvegicus*) and black (*Rattus rattus*) rats on their social behavior. *Doklady Biological Sciences* (2007b) 416(1):368-370.

Dzhemukhadze N K, Kiladze A B. Comparison of the activity of some phosphatases in the midventral gland and nonspecific sebaceous glands of the neck in the Campbell hamster (*Phodopus campbelli*). *Doklady Biological Sciences* (2008) 423(1):447-449.

Dzhemukhadze N K, Kiladze A B. A multilevel model for chemocommunication of two Lasiopodomys species based on histoenzymatic activity of skin glands. *Uspekhi sovremennoi biologii* (2012) [*Success in modern biology*] 132(2):200-210. (In Russian)

Dzhemukhadze N K, Kiladze A B. Histoenzymatic activity of the skin glands of rats: Chemocommunication aspect. *Glavnyj zootekhnik* (2013) [*chief zootechnologist*] 5:42-47. (In Russian)

Dzhemukhadze N K, Kiladze A B. *Histochemistry of phosphatases of skin glands in a comparative series of voles of the subfamily Arvicolinae.* Saarbrücken: LAP Lambert Academic Publishing; 2014. (In Russian)

Everitt B S, Landau S, Leese M. *Cluster Analysis.* 4th ed. New York, NY: Oxford University Press, Inc.; London: Arnold; 2001.

Feng A Y T, Himsworth C G. The secret life of the city rat: a review of the ecology of urban Norway and black rats (*Rattus norvegicus* and *Rattus rattus*). *Urban Ecosystems* (2014) 17(1):149-162.

Fisher R A. The use of multiple measurements in taxonomic problems. *Annals of Eugenics.* (1936) 7(2):179-188.

Gomori G. *Microscopic histochemistry: Principles and practice*. Chicago, IL: University of Chicago Press; 1952.

Gower J C. A comparison of some methods of cluster analysis. *Biometrics* (1967) 23(4):623-637.

Haniza M Z H, Adams S, Jones E P, MacNicoll A, Mallon E B, Smith R H, Lambert M S. Large-scale structure of brown rat (*Rattus norvegicus*) populations in England: effects on rodenticide resistance. *Peer. J.* (2015) 3:e1458.

Harbison B. PMPs re-strategize rodent control in response to COVID-19 pandemic. *Pest Control Technology*. April 16, 2020. Available from: https://www.pctonline.com/article/pmps-restrategize-rodent-control-respose-covid-19/. Accessed on 22 April 2020.

Hashimoto K, Ogawa K. Histochemical studies on the skin. I. The activity of phosphatases during the histogenesis of the skin in the rat. *American Journal of Anatomy* (1963) 113:35-50.

Hedrich H J. History, strains and models. In: Krinke G J (ed) *The Laboratory Rat*. San Diego, CA: Academic Press; 2000. pp. 3-16.

Hrytsevych N R, Vereschaka V V. Features of the ultrastructure of the skin of white rats 60 and 90 days after modeling of portal hypertension. *Wiadomości Lekarskie* (2021). 74(9cz1):2197-2201.

Khalafyan A A. STATISTICA 6. *Statistical data analysis*. Moscow: Binom-Press; 2008. (In Russian)

Kiladze A B, Dzhemukhadze N K. *Qualimetry in the histochemistry of enzymes in mammalian skin glands*. Moscow: Infra-Inzheneriya [infra-engeneering]; 2013. (In Russian).

Kiladze A B, Dzhemukhadze N K. *Histoenzymatic model of chemocommunication of voles of the subfamily Arvicolinae (based on the activity of phosphatases of the skin glands)*. Moscow: Ru-Science; 2016. (In Russian)

Kiladze A B, Dzhemukhadze N K. Evaluation of sexual dimorphism of histochemical activity of phosphatases of the plantar glands of Norway rats (*Rattus norvegicus*). *Biosystems Diversity*. (2019) 27(1):39-42.

Kiladze A B, Dzhemukhadze N K. *Bioqualimetric study of the phosphatases activity of skin glands of Norway rats and black rats*. Moscow/Izhevsk: Institute of *Computer Science*; 2020. (In Russian)

Kiladze A B, Dzhemukhadze N K. Bioinformatics model of histoenzymatic parameters of the skin glands of Norway (*Rattus norvegicus*) and black (*Rattus rattus*) rats. *Ukrainian Journal of Ecology*. (2021a) 11(5):51-65.

Kiladze A B, Dzhemukhadze N K. Bioinformatics analysis of the sexual dimorphism of the skin glands histoenzymatic activity in Norway Rats (*Rattus norvegicus*). In: Berhardt LV (ed) *Advances in Medicine and Biology*. Volume 191. Hauppauge, NY: Nova Science Publishers; 2021b. pp. 175-188.

Kiladze A B, Dzhemukhadze N K. Bioinformatics of sexual dimorphism of histoenzymatic activity in the skin glands of laboratory rats. *JP Journal of Biostatistics* (2021c) 18 (3):429-436.

Kiladze A, Dzhemukhadze N. Bioinformatics analysis of sexual dimorphism of histochemical activity of the skin glands of field vole. *Veterinariya sel'skohozyajstvennyh zhivotnyh* [*Veterinary medicine of farm anymals*] (2022a) 1:35-41. (In Russian)

Kiladze A B, Dzhemukhadze N K. Bioinformatics of seasonal indicators of phosphatase activity of nonspecific skin glands of Campbell's dwarf hamster. *Conference "Mammals in a changing world: Actual problems of theriology"* (XI Congress of the Theriological Society at the Russian Academy of Sciences). Moscow: KMK Scientific Press; 2022b. p. 149. (In Russian)

King C M, Foster S, Miller S. Invasive European rats in Britain and New Zealand: same species, different outcomes. *Journal of Zoology* (2011) 285(3):172-179.

Klemann N, Walther B, Matuschka F R, Jacob J, Endepols S. The stop-feed effect of cholecalciferol (vitamin D_3) and the efficacy of brodifacoum combined with cholecalciferol in Y139C-resistant Norway rats (*Rattus norvegicus*). *Journal of Pest Science* (2023) 2004:1-10.

Lobanov A S. The basic concepts of qualimetry. *Scientific and Technical Information Processing* (2013) 40:72-82.

Lojda Z, Gossrau R, Schiebler T H. *Enzyme histochemistry: A laboratory manual*. Berlin/ Heidelberg: Springer-Verlag; 1979.

Luppa H. *Basic histochemistry*. Moscow: Mir; 1980. (In Russian)

Marcoulides G A, Hershberger S L. *Multivariate statistical methods: A first course*. New York, NY: Psychology Press; 1997.

Marusina M Ya, Tkalich V L, Labkovskaya R Ya. *Metrological support of measuring instruments*. St. Petersburg: ITMO University; 2019. (In Russian)

Mazzotta E, Bellinati L, Bertasio C, Boniotti M B, Lucchese L, Ceglie L, Martignago F, Leopardi S, Natale A. Synanthropic and wild animals as sentinels of zoonotic agents: A study of Leptospira genotypes circulating in Northeastern Italy. *Int. J. Environ. Res. Public Health* (2023) *20*:3783.

McLachlan G J. *Discriminant analysis and statistical pattern recognition*. Hoboken: John Wiley & Sons; 2004.

Meier-Ruge W A, Bruder E. Current concepts of enzyme histochemistry in modern pathology. *Pathobiology* (2008) 75:233-243.

Montagna W, Noback C R. Histochemical observations on the sebaceous glands of the rat. *American Journal of Anatomy* (1947) 81:39-61.

Morrison D F. *Multivariate statistical methods*. London: Duxbury Press; 2004.

Nachlas M M, Crawford D T., Goldstein T P, Seligman A M. The histochemical demonstration of cytochrome oxidase with a new reagent for the Nadi reaction. *Journal of Histochemistry & Cytochemistry* (1958) 6:445-456.

Niczyporuk M. Rat skin as an experimental model in medicine. *Progress in Health Sciences* (2018) 8(2):223-228.

Parsons M H, Kiyokawa Y, Richardson J L, Stryjek R, Byers K A, Himsworth C G, Corrigan R M, Deutsch M A, Ootaki M, Tanikawa T, Parsons F E, Munshi-South J. Rats and the COVID-19 pandemic: Early data on the global emergence of rats in response to social distancing. *MedRxiv* (2020).

Pearse A G E. *Histochemistry: Theoretical and Applied*. 2nd ed. London: J. and A. Churchill Ltd.; 1960.

Pritchett-Corning K. Rat. In: Liss C, Litwak K, Tilford D, Reinhardt V (eds) *Comfortable Quarters for Laboratory Animals.* Tenth Edition. Washington, DC: Animal Welfare Institute; 2015. pp. 20-37.

Rozhkov N N. *Qualimetry and quality management. Mathematical methods and models.* Moscow: Yurayt Publishing House; 2018. (In Russian)

Saga K. Histochemical and immunohistochemical markers for human eccrine and apocrine sweat glands: an aid for histopathologic differentiation of sweat gland tumors. *Journal of Investigative Dermatology Symposium Proceedings* (2001) 6(1):49-53.

Saga K. Structure and function of human sweat glands studied with histochemistry and cytochemistry. *Progress in Histochemistry and Cytochemistry* (2002) 37(4):323-386.

Saga K, Morimoto Y. Ultrastructural localization of alkaline phosphatase activity in human eccrine and apocrine sweat glands. *Journal of Histochemistry and Cytochemistry* (1995) 43(9):927-932.

Schweinfurth M K. The social life of Norway rats (*Rattus norvegicus*). *eLife* (2020) 9:e54020.

Searle A, Turnbull J, Lorimer J. After the anthropause: Lockdown lessons for more than human geographies. *The Geographical Journal* (2021) 187:69-77.

Sharp P, Villano J S. *The laboratory rat.* Boca Raton, FL: CRC Press; 2012.

Soh S, Chua C H, Neo Z W, Kong M, Ong B L, Aik J. Rodent activity in municipal waste collection premises in Singapore: an analysis of risk factors using mixed-effects modelling. *Sci. Rep.* (2023) 13(1):3070.

Sokolov V. *Mammal skin.* Berkeley, CA: University of California Press; 1982.

Sokolov V E, Chernova O F. *Cutaneous glands of mammals.* Moscow: GEOS; 2001. (In Russian)

Sokolov V E, Dzhemukhadze N K. Comparative histochemistry of enzymes in various parts of the skin of some members of the rodent order (Rodentia). *Chemical signals of animals.* Moscow: Nauka; 1982. (In Russian)

Sokolov V E, Skurat L N, Stepanova L V, Sumina E B, Shabadash S A. *Guide for analysis of skin of mammals.* Moscow: Nauka; 1988. (In Russian)

Stevens S S. On the theory of scales of measurement. *Science.* New Series (1946) 103(2684):677-680.

Uea-Anuwong T, Byers K A, Wahl L C, Nekouei O, Grohn Y T, Magouras I. Antimicrobial resistance in bacteria isolated from peridomestic Rattus species: A scoping literature review. *One Health* (2023) 16:100522.

Warburton A L, Santer R M. Sympathetic and sensory innervation of the urinary tract in young adult and aged rats: a semi-quantitative histochemical and immuno-histochemical study. *The Histochemical Journal* (1994) 26(2):127-133.

Wick M R. Histochemistry as a tool in morphological analysis: a historical review. *Annals of Diagnostic Pathology* (2012) 16(1):71-78.

Wilks S S. Some aspects of quantification in science. *Isis* (1961) 52(2):135-142.

Chapter 6

The Health Benefits of Fucoidan in Improving Gastrointestinal Function

Aparna Kannan
and Anisha G. S.*
Post-Graduate and Research Department of Zoology, Government College for Women, Thiruvananthapuram, Kerala, India

Abstract

Fucoidan polysaccharide has tremendous potential to improve the gastrointestinal function of humans and animals by way of modulating the gut microbiome and regulating the immune response. Owing to its complex structure, fucoidan can circumvent the digestive action of gastric and pancreatic digestive enzymes and hence becomes available as a prebiotic for fermentation by the gut microbiota. Several researchers have reported that fucoidan promotes the growth of beneficial bacteria such as Lactobacillus sp., Bifidobacterium longum and B. bifidum over the pathogenic bacteria of the genera Peptococcus. Dietary supplementation of fucoidan can modulate and reorganize the gut microbiota and thereby influence the inflammatory response in the intestine. Fucoidans are also useful as chemotherapeutics or adjuvants to chemotherapy in the treatment of gastrointestinal cancers. Fucoidans from different algal sources differ in their monosaccharide composition, degree of polymerization and sulfation, which are the decisive factors in their bioactive properties. It is plausible to propose fucoidan as a promising nutraceutical to improve gastrointestinal function.

* Corresponding Author's Emails: anishags@gcwtvm.ac.in, anisgrace@gmail.com, anisgrace@yahoo.co.in

In: Advances in Biology. Volume 4
Editor: Charles D. Grant
ISBN: 979-8-88697-999-2

Keywords: colon cancer, fucoidan, functional food, gut microbiota, inflammatory response, intestinal dysbiosis, prebiotic

Introduction

The marine brown macroalgae belonging to the Class Phaeophyceae contain copious amounts of fucoidan, the sulfated polysaccharide with excellent bioactive properties such as immunomodulation, antioxidant, antibacterial and antiviral actions, to mention a few. Additionally, fucoidan has numerous health benefits that help improve gastrointestinal function in humans and animals. Low-molecular-weight fucoidans have more bioactivity than their high-molecular-weight counterparts (Hwang et al., 2016). The low-molecular-weight fucoidans can be absorbed into the blood circulation and distributed to cells (Flórez-Méndez and González 2019), whereas high-molecular-weight fucoidans have difficulty crossing the membrane barriers of the cells. However, the high-molecular-weight fucoidans can escape the digestive action of the enzymes and gastrointestinal secretions and reach the lower gastrointestinal tract where they can be acted upon by the inhabitant probiotic microflora. The metabolic products produced by the microbial action on these polysaccharides can influence the immunity of the host (Yuan et al., 2021).

The gastrointestinal tract is always exposed to microbes which may include beneficial microbes or potential pathogens (Hwang et al., 2016). Maintenance of gastrointestinal health requires that the probiotic bacterial populations are favoured over pathogenic bacteria. The immune system cells present in mucosal-associated lymphoid tissues and gut-associated lymphoid tissues are watchful in safeguarding the gastrointestinal tract from the attack of pathogenic microorganisms. Bioactive polysaccharides of macroalgal origin are recognized to have beneficial effects in modulating the immune function of the gastrointestinal tract and maintaining good bowel health. Seaweeds have long been included in the traditional cuisines of Asia owing to their nutritional value. Fucoidan polysaccharides from marine brown macroalgae form part of soluble dietary fibre (Flórez-Méndez and González 2019; Rajapakse and Kim 2011) and have remarkable immunomodulatory action, antibacterial action and prebiotic potential which are strategic in maintaining the healthy status of the gastrointestinal function. The bioactive properties of fucoidan advocate their use as functional food ingredients to equip consumers against several physiological disorders and illnesses. Bioactive compounds have advantageous effects on health owing to their ability to influence the physiological and cellular processes even if present in small concentrations in the food (Flórez-Méndez and González 2019). Fucoidan is anionic because of the presence of negatively charged sulfate groups present in its structure. The anionic groups can associate with oppositely charged chemical groups in pharmaceutical formulations. Moreover, fucoidan is less toxic, easily biodegradable and biocompatible to be used as functional food ingredients (Liyanage, Nagahawatta, and Jayawardena 2023).

Structure of Fucoidan and Macroalgal Sources

Since Kylin first isolated fucoidan in 1913, the structures of fucoidans from different brown sea weeds have been investigated. The sea weed fucoidans are heterogenic that means they represent the mixture of structurally related polysaccharides with certain variations of the content of carbohydrate units and non-carbohydrate substituents. Figure 1 shows the structure of fucoidan. Fucoidan are mainly composed of fucose and sulphate, also contain other monosaccharides like mannose, galactose, glucose, xylose etc. The type I chains are found to contain the repeating (1→3) linked alpha L fucopyranose residues, whereas type II chains contain the alternating (1→30 and (1→4) may occur at positions 2, 3, and 4 and the monosaccharides are associated via alpha-1,2, alpha-1,3 or alpha-1,4 glycosidic bonds. Table 1 enlists the brown macroalgae producing fucoidan and their structural backbone.

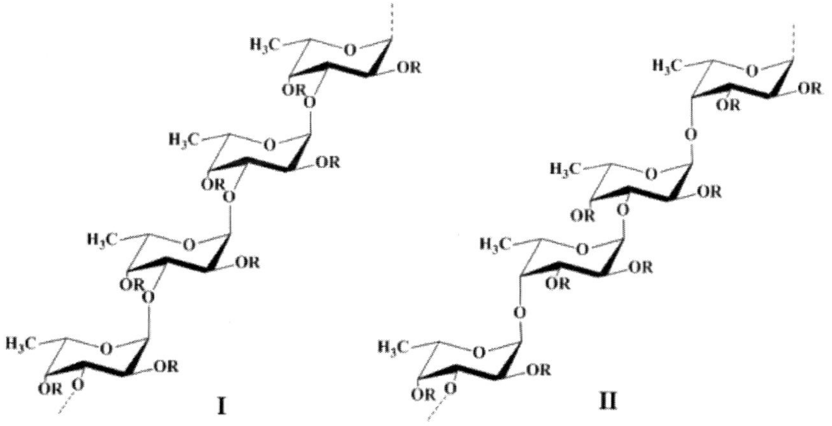

Figure 1. Structure of fucoidan.

Table 1. Brown macroalgal sources of fucoidan

Brown macroalgal sources	Order	Backbone
Fucus vesiculosus	Fucales	α(1→3) and α(1→4) linked fucose
Ascophyllum nodosum	Fucales	α(1→3) and α(1→4) linked fucose
Fucus evanescens	Fucales	α(1→3) and α(1→4) linked fucose
Fucus serratus	Fucales	α(1→3) and α(1→4) linked fucose
Sargassum linifolium	Fucales	a backbone chain of D-glucuronic acid and D-mannose residues
Sargassum stenophyllum	Fucales	α(1→3) and α(1→4) linked fucose
Sargassum fusiforme	Fucales	α(1→2) linked α-d-Manp and α(1→4) linked β-d-GlcpA
Chorda filum	Laminariales	α(1→2) and α(1→3) linked fucose
Laminaria saccharina	Laminariales	α(1→2) and α(1→3) linked fucose
Lessonia vadosa	Laminariales	α(1→3) linked fucose
Cladosiphon okamuranus	Chordariales	α(1→2) and α(1→3) linked fucose
Adenocystis utricularis	Ectocarpales	mainly composed of 3-linked α-l-fucopyranosyl backbone
Analipus japonicas	Ectocarpales	mainly α(1→3) linked fucose

Role of Fucoidan in Improving Gastrointestinal Function

Brown algae have been included in traditional cuisines considering their health benefits. The health benefits of fucoidan in gastrointestinal function are associated with their bioactive properties such as antioxidant, anti-inflammatory, anti-apoptotic, anti-obesogenic, anti-hyperlipidemic and anti-diabetic actions. Figure 2 depicts the beneficial bioactive properties of fucoidan in improving gut health.

Prebiotic Potential of Fucoidan

The beneficial bacteria inhabiting the gastrointestinal lining include the members of the genera *Bifidobacteria*, *Lactobacillus,* etc. These beneficial bacteria are present in fermented food and are consumed worldwide as probiotic bacteria. The probiotic bacteria have the advantage of promoting the immunity of the hosts by preventing the pathogenic bacteria from adhering to the gastrointestinal epithelium. The term gut microbiota includes not only bacteria but also all microorganisms such as fungi, protists, archaea and viruses inhabiting the gastrointestinal tract (Hrncir 2022). Many of these microbes are opportunistic pathogens. Hence maintaining a healthy balance between the probiotic bacteria and other microorganisms is essential for the host immune system, or else it may result in intestinal dysbiosis, a condition in which the gastrointestinal tract becomes overloaded with microbiota that can adversely affect the host immune system and metabolism (Hrncir 2022). The polysaccharides that can promote the adhesion and growth of probiotic bacteria in the gastrointestinal epithelium can act as prebiotics that prevents intestinal dysbiosis and help maintain an ideal bacteria profile for healthy intestinal barrier function (Hwang et al., 2016).

Promoting Colonization of Beneficial Bacteria

Fucoidan is projected as a potential prebiotic agent in functional food which can selectively promote the adhesion and growth of beneficial bacteria and boost the host's natural immunity against opportunistic pathogenic bacteria (Shang et al., 2016; Hwang et al., 2016). Using high throughput sequencing and bioinformatics analysis, Shang et al., (2016) found out that fucoidan treatment could modulate the gut microbiota in mice and the abundance of beneficial bacteria such as *Lactobacillus* and *Ruminococcaceae* could be achieved at the expense of the opportunistic pathogenic bacteria *Peptococcus*. The Caco2 cell line co-cultured with *Bifidobacterium lactis* showed a dose-dependent promotion of bacterial growth with fucoidan treatment ranging from 10 to 300 μg/mL (Hwang et al., 2016) (Table 2). Fucoidan also enhanced stronger cell adhesion of *B. lactis* to Caco2 cells in culture.

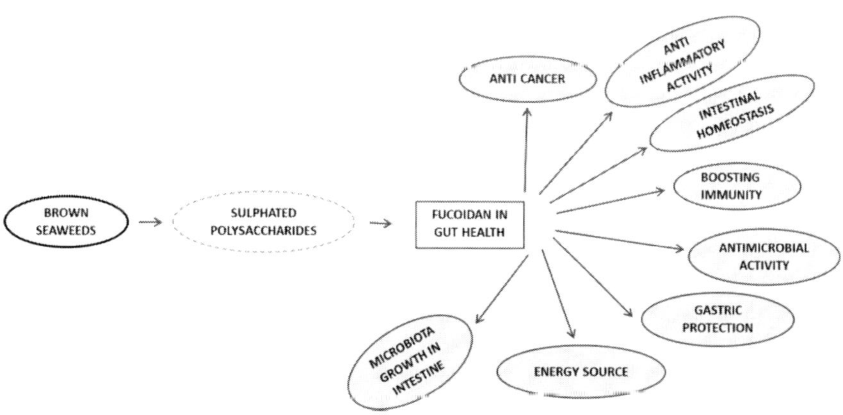

Figure 2. Beneficial bioactive properties of fucoidan in improving gut health.

Table 2. Bioactive properties of fucoidan and their implications in improving gastrointestinal function

Macroalgal source of fucoidan	Study model	Bioactive properties benefiting gastrointestinal function	Reference
Sargassum hemiphyllum	Caco2 cell line co-culture with *Bifidobacterium lactis* treated with fucoidan + fucoxanthin	Activate probiotic growth Reduce the inflammation of the intestinal epithelial cells Enhance intestinal epithelial barrier and immune function against the lipopolysaccharide effect Inhibit IL-1β and TNF-α Promote IL-10 and IFN-γ	(Hwang et al., 2016)
Ascophyllum nodosum	C57BL/6J mice treated with ciprofloxacin-metronidazole and fucoidan	Prevent colon shortening Alleviated colonic tissue damages Reverse the alteration of gut microbiota by increasing potentially beneficial bacteria (Ruminococcaceae_UCG_014 and Akkermansia) and decreasing the harmful bacteria, (Proteus and Enterococcus) Suppress overproduction of TNF-α, IL-1β, and IL-6 and promote IL-10. Regain the production of short-chain fatty acids	(Wang et al., 2020)
Commercial fucoidan	SPF C57BL/6J mice fed with fucoidan diet	Enrichment of *Lactobacillus animalis*, *Lactobacillus gasseri* and *Lactobacillus reuteri* population in gut Decrease in serum triglycerides and LPS-binding proteins Increase in tryptophan related metabolites (indoleacrylic acid, 3-indoleacrylic acid, and 5-hydroxytryptamine)	(Yuan et al., 2021)
Laminaria japonica and *Ascophyllum nodosum*	Mice with high-fat-diet-induced metabolic syndrome	Alleviate diet-induced metabolic syndrome (MetS), Reduce body weight, fasting blood glucose, hepatic steatosis and systemic inflammation Enrichment of beneficial microbes (Akkermansia muciniphila, Alloprevotella, Blautia and Bacteroides)	(Shang et al., 2017)

Macroalgal source of fucoidan	Study model	Bioactive properties benefiting gastrointestinal function	Reference
Fucus vesiculosus	IEC-6 cells (in vitro study) Db/db mice (in vivo study)	Decrease postprandial hyperglycemia Inhibit α-glucosidase activity (possible treatment of type 2 diabetes) Decrease fasting blood glucose and hemoglobin A1c (HbA1c) levels of db/db mice	(Shan et al., 2016)
Laminaria japonica	Streptozotocin-induced diabetic mice (C57BL/6)	Alleviate diabetic retinal neovascularization and Decrease HIF-1α and VEGF expressions (potential for treatment of diabetic retinopathy)	(Yang et al., 2013)
Saccharina japonica	BALB/c mice	Increase in mRNA expression of galectin-9 in intestinal epithelial cells and its serum levels Suppression of anaphylaxis reaction in food allergy	(Tanino et al., 2016)
Undaria pinnatifida	Genetically diabetic C57BL/KSJ db/db female mice and their age-matched non-diabetic littermates C57BL/KSJ m+/+db	Prevention of hyperglycemia in non-diabetic mice Reduction of blood glucose levels in diabetic mice	(Kim, Yoon, and Lee 2012)
Saccharina japonica	Male Wistar rats	Alleviate diabetic nephropathy Decrease blood glucose, blood urea nitrogen, creatinine clearance and urine creatinine Increase microalbumin, serum insulin and β2-microglobulin	(Wang et al., 2014)
Laminaria japonica	Goto-Kakizaki type 2 diabetic rats and age-matched Wistar rats	Increase the phosphorylation of endothelial nitric oxide synthase and NO synthesis Improve endothelium-dependent vasodilation Reduce diabetes-induced hypertension	(Cui et al., 2014)
Ascophyllum nodosum	Weaned pigs naturally infected with Salmonella	Reduction of Salmonella typhimurium counts Decrease the expression of interleukin 22 and TGF-1β in the ileum	(Venardou et al., 2022)

Fucoidan is also highly useful in the reclamation of the adequate bacteria profile in the gut microbiota following an antibiotic treatment that causes serious damage to the intestinal mucosal lining and the gut microbiome. Dietary administration of fucoidan isolated from *Ascophyllum nodosum* to C57BL/6J mice following ciprofloxacin-metronidazole treatment was demonstrated to recuperate the microbial composition by increasing the growth of beneficial bacteria such as *Ruminococcaceae*_UCG_014 and *Akkermansia* while decreasing the populations of harmful bacteria such as *Proteus* and *Enterococcus* (Wang et al., 2020) (Table 2).

Alleviation of Endotoxemia and Intestinal Dysbiosis

The Gram-negative bacteria in the gut microbiome synthesize the endotoxin Slipopolysaccharides (LPS) which reach the systemic circulation and cause endotoxemia (Yuan et al., 2021). The serum lipopolysaccharide-binding proteins are the biomarkers of endotoxemia. Fucoidan treatment can alleviate intestinal dysbiosis and help establish a favourably balanced profile in the gut microbiota which would considerably decrease the antigen load in the intestinal tract. The decrease in the antigen load correspondingly lowers the serum lipopolysaccharide-binding protein levels indicating attenuation of endotoxemia (Shang et al., 2016; Yuan et al., 2021).

Influencing Metabolomics of the Host

The abundance of beneficial microbes established in the gastrointestinal tract by fucoidan treatment is significant in influencing the metabolomics of the host, such as serum levels of lipid metabolites and aminoacid-associated metabolites. The SPF C57BL/6J mice administered with fucoidan for 10 weeks showed an abundance of *Lactobacillus animalis*, *Lactobacillus gasseri* and *Lactobacillus reuteri* which was correlated with a decrease in serum triglyceride levels and an increase in tryptophan-associated metabolites such as indoleacrylic acid, 3-indoleacrylic acid and 5-hydroxytryptamine (Yuan et al., 2021).

Alleviation of Metabolic Syndrome

Metabolic syndrome is a metabolic disorder characterized by physiological conditions such as obesity, insulin resistance, dyslipidemia and hypertension (Shang et al., 2017). Metabolic syndrome can be induced by a high-fat diet that alters the microbial profile in the gut microbiota leading to intestinal dysbiosis which can in turn modulate the metabolism and immunity of the host.

Dietary administration of fucoidan can attenuate metabolic syndrome by direct and indirect mechanisms. The direct mechanisms involve the direct intervention of fucoidan in different pathways associated with metabolism, whereas indirect mechanisms involve promoting the colonization of gastrointestinal epithelium with beneficial microbes which in turn exert beneficial actions that alleviate the metabolic disorders of the host. Fucoidan can exert a hypoglycemic effect directly by inhibiting the α-glucosidase enzyme (Shan et al., 2016) and increasing the serum levels of insulin (Mukhamejanov et al., 2019) suggesting a possible therapeutic approach for the treatment of type 2 diabetes mellitus. Fucoidan also interferes with the vascular endothelial growth factor (VEGF) signaling pathway that has implications for diabetic retinopathy, the leading cause of blindness in diabetic patients worldwide (Yang et al., 2013). Diabetic retinopathy occurs due to microvascular dysfunction and neovascularization in the retina which is triggered by VGEF and hypoxia-inducible factor-1α (HIF-1α). Fucoidan can lower the expression of HIF-1α and VEGF thereby preventing microvascular dysfunction and neovascularization of the retina (Yang et al., 2013). Fucoidan has also been reported as effective in alleviating diabetic nephropathy (Wang et al., 2014). The metabolic disorder hyperglycemia is the primary risk factor associated with the onset of diabetic nephropathy. The mechanism of preventive or alleviating action of fucoidan in diabetic nephropathy is related to its antioxidant property. Hyperglycemia in diabetes triggers an overproduction of reactive oxygen species (ROS) succumbing the cells to oxidative stress. Fucoidan can either directly act as an exogenous antioxidant or enhance the activity of endogenous antioxidants reducing the levels of ROS and alleviating the manifestations of diabetic nephropathy and chronic kidney disease (Ma et al., 2022). Fucoidan reduces oxidative stress in chronic kidney disease through GSK3β-Nrf2-HO-1 signaling pathway (Ma et al., 2022). The male Wistar rats administered with fucoidan from *Saccharina japonica* demonstrated a decrease in blood glucose, blood urea nitrogen, creatinine clearance and urine creatinine and a significant increase in microalbumin, serum insulin and β2-microglobulin (Wang et al., 2014). Another example of the direct intervention of fucoidan in alleviating metabolic syndrome is the improvement of insulin sensitivity and restoration of lipid homeostasis through the activation of the AMP-activated protein kinase (AMPK) signaling pathway (Jeong et al., 2013). Diabetes mellitus is associated with cardiovascular pathologies characterized by endothelial dysfunction due to impairment of endothelial nitric oxide synthase (eNOS) and nitric oxide bioavailability (Cui et al., 2014). Treatment with low-molecular-weight fucoidan can reverse the manifestations of diabetes-induced cardiovascular pathologies. Low-molecular-weight fucoidan can dose-dependently increase the phosphorylation of eNOS and NO synthesis thereby improving endothelium-dependent vasodilation and reducing diabetes-induced hypertension (Cui et al., 2014).

Fucoidan-based diet can indirectly attenuate metabolic syndrome by promoting the colonization of beneficial microbes such as *Akkermansia muciniphila*, *Alloprevotella*, *Blautia* and *Bacteroides* (Shang et al., 2017). The LPS produced by potentially harmful bacteria such as *Enterobacter cloacae* can instigate insulin resistance and obesity by eliciting the inflammation-mediated pathway (Fei and Zhao 2013). Populating the intestinal epithelium by microbes such as *A. muciniphila* promotes glucose tolerance and reduces the chances of obesity even on a high-fat diet (Zhou 2017; Shang et al., 2017). Additionally, reducing the number of undesirable *Enterobacter* populations can correspondingly reduce the body weight of the host alleviating the difficulties associated with obesity such as hyperglycemia and hypertension (Fei and Zhao 2013). *A. muciniphila* helps in maintaining the integrity of the gut barrier (Zhou 2017), whereas *Alloprevotella*, *Blautia* and *Bacteroides* produce short-chain fatty acids (Shang et al., 2017). The short-chain fatty acids help create a congenial microenvironment in the gut for the growth of microbes (Luthuli et al., 2019). They also improve the epithelial barrier

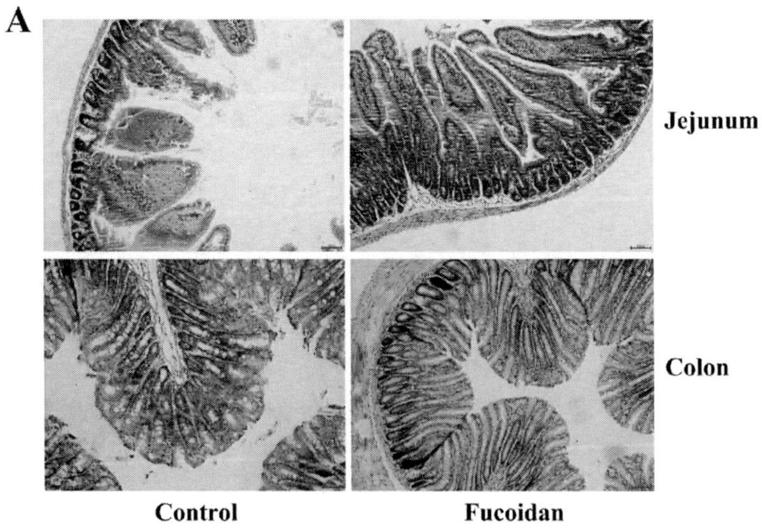

Figure 3. (Continued).

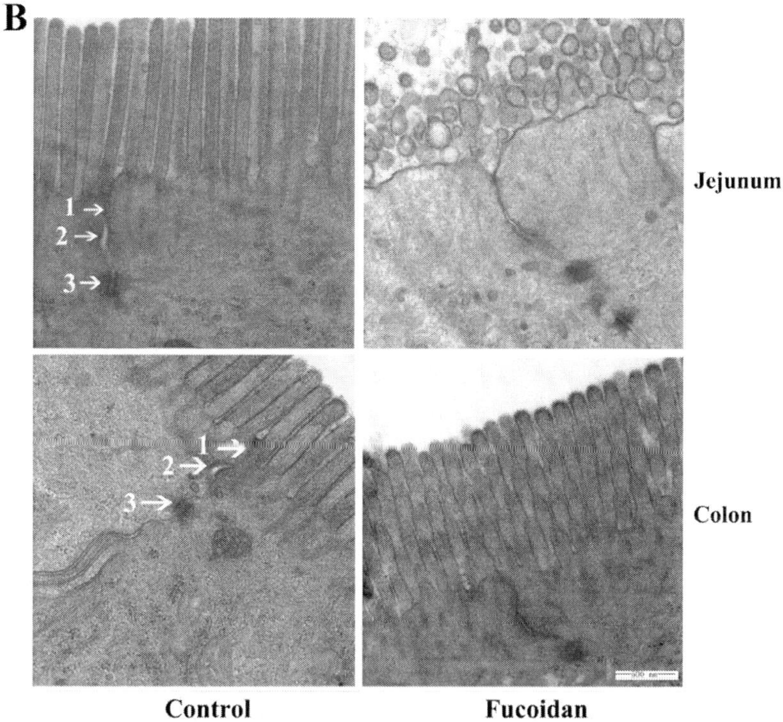

Figure 3. Pathological changes of the jejunal and colonic tissues of non-diabetic (NOD) mice. (A) Histopathological changes of intestinal epithelium revealed under optical microscope by hematoxylin and eosin staining. The tissue from the control group exhibit histological damage with increased infiltration of inflammatory cells. However, fucoidan attenuated the pathological changes in the jejunal and colonic tissues of NOD mice. (B) Ultrastructure observations of intestinal tissues under Transmission Electron Microscope. The tight junctions are fuzzy and slightly relaxed with low electron density in the control group. In the fucoidan group, histological morphology and the tight junctions are improved. 1, tight junction; 2, adherens junction; 3, desmosome. (Sun et al., 2020) (http://creativecommons.org/licenses/by-nc-nd/4.0/)

function and cardiovascular health (Ricós-Muñoz, Maicas, and Pina-Pérez 2021). Enrichment of the beneficial microbes following fucoidan treatment decreases the body weight, fasting blood sugar and liver steatosis in the host indicating the anti-obesogenic and anti-diabetic actions of fucoidan that can alleviate metabolic syndrome (Shang et al., 2017; Yang and Lim 2021).

Improving Intestinal Epithelial Barrier Function

An efficient intestinal epithelial barrier develops from the normal colonization of bacterial populations. Transepithelial resistance indicating the tightness of the tight junctions between the intestinal epithelial cells (Hwang et al., 2016), is decisive in the efficiency status of the intestinal epithelial barrier. Highly efficient tight junctions prevent bacterial invasion across the intestinal epithelial barrier and entry into the systemic circulation. Poor efficiency of tight junctions may deteriorate the integrity and increase the permeability of the intestinal epithelial barrier giving way to pathogens to invade the systemic circulation and activate the autoimmune system. Increased permeability of the intestinal epithelial barrier and activation of the autoimmune response are reported to precede the onset of autoimmune disorders such as type 1 diabetes mellitus which occurs due to the destruction of β-cells in the pancreatic islets of Langerhans (Lee et al., 2010).

Integral membrane proteins in the plasma membrane such as occludin, claudin-1, claudin-2, zonulin and scaffold proteins ZO-1 form part of the tight junctions that act as intercellular gates in the intestinal epithelial barrier (Hwang et al., 2016; Sun et al., 2020). The Gram-negative bacteria in the gut microbiome synthesize lipopolysaccharides (LPS) that deteriorate transepithelial resistance by inhibiting the mRNA expression of occludin, claudin-1 and claudin-2. The intact and non-disrupted intestinal barrier prevents the infiltration of bacterial endotoxins and diamine oxidase (DAO) and D-lactic acid (D-LA) from reaching the circulation (Sun et al., 2020). The presence of LPS, DAO and D-LA in the circulation is an indication of intestinal epithelial hyperpermeability. On the contrary, fucoidan treatment can recover the mRNA expression of the tight junction proteins (occludin, claudin-1, claudin-2, zonulin and ZO-1) while decreasing LPS, DAO and D-LA in the blood and improve transepithelial resistance thus enhancing intestinal epithelial barrier function (Hwang et al., 2016; Sun et al., 2020). Maintaining the integrity of the intestinal epithelial barrier is proposed to be essential for preventing the onset of type 1 diabetes mellitus. Figure 3 reveals the effect of fucoidan treatment in maintaining the integrity of the intestinal epithelial barrier.

Anti-Inflammatory Action in the Gastrointestinal Lining

The LPS produced by Gram-negative bacteria elicits an inflammatory immune response in the gastrointestinal epithelium leading to local and systemic organ dysfunction (Hwang et al., 2016). The elicitation of the inflammatory response by the intestinal epithelium is marked by the release of pro-inflammatory cytokines such as interleukin-1β (IL-1β) and tumor necrosis factor-α (TNF-α) produced by the macrophages or monocytes. The excess inflammatory response is regulated by the anti-inflammatory cytokines interleukin-10 (IL-10) and interferon-γ (IFN-γ).

The anti-inflammatory action of fucoidan is effective in inhibiting LPS-induced damage to intestinal epithelial cells and increasing cell viability. Fucoidan reduces inflammatory response by decreasing the release of IL-1β and TNF-α. This is mediated through the inhibition of the transcription regulator nuclear factor kappa B (NF-κB) (Hwang et al., 2016). Moreover, fucoidan also promotes the secretion of anti-inflammatory IL-10 and IFN-γ. Fucoidan synergistically improves the anti-inflammatory activity when used in combination with other anti-inflammatory compounds such as fucoxanthin (Hwang et al., 2016).

Antibiotic treatment may induce side effects such as inflammation of the intestinal mucosal lining. The anti-inflammatory action of fucoidan is beneficial in relieving the side effects of antibiotic treatment. Fucoidan from *A. nodosum* administered to ciprofloxacin-metronidazole-treated C57BL/6J mice showed good anti-inflammatory action by suppressing the excess production of inflammatory cytokines such as TNF-α, IL-1β, and IL-6 and promoting the expression of anti-inflammatory cytokine IL-10 (Wang et al., 2020).

Fucoidan can prevent the transendothelial transmission of leucocytes by binding with them and slowing down their movement to the endothelium surface (Liyanage, Nagahawatta, and Jayawardena 2023). Fucoidan also alleviates the symptoms associated with the major inflammatory diseases occurring in the gastrointestinal tract such as ulcerative colitis (UC) and Crohn's disease (CD) (Yang and Lim 2021).

Prevention of Food Allergies

Oral administration of fucoidan is reported to provide anti-allergic effects. Fucoidan is effective against type I allergy to prevent passive cutaneous anaphylaxis reactions (Tanino et al., 2016). The mechanism of the anti-allergic effect of fucoidan proposed by Tanino et al., (2016) illustrates that fucoidan from *Saccharina japonica* induces the mRNA expression of galectin-9 in the intestinal epithelial cells and its release into the blood (Table 2). Galectin-9 is a member of the soluble lectin family and is associated with the prevention of autoimmune diseases and food allergies. The galectin-9 induced by fucoidan administration binds with the immunoglobulin E (IgE) thus interfering with the binding of IgE to IgE receptor FcεRI on mast cells. Thus galectin-9 blocks the formation of the IgE-antigen complex (Niki et al., 2009). This prevents mast cell activation and suppresses anaphylaxis reactions. The heightened expression of galectin-9 is effective in suppressing food allergies. The oral administration of fucoidan is more effective in preventing passive cutaneous anaphylaxis reaction than its intraperitoneal administration.

Maintenance of Colon Health and Prevention of Colorectal Cancer

Fucoidan in brown algae can act as dietary fibres in the gastrointestinal tract. They are metabolized by the colon bacteria to produce short-chain fatty acids which have a significant impact on maintaining the good condition of the colon. The short-chain fatty acids produced by the colon bacteria are reported to have anti-cancer and anti-inflammatory effects (Cotas et al., 2021). Dietary fucoidan helps prevent colon shortening and alleviate colon tissue damage following antibiotic treatment. The C57BL/6J mice after ciprofloxacin-metronidazole treatment showed revitalization and maintenance of colon health on dietary administration of fucoidan from *A. nodosum* (Wang et al., 2020). The ciprofloxacin-metronidazole-treated mice also showed a regain of short-chain fatty acid synthesis.

Fucoidan has anticancer property which is beneficial in the treatment and prevention of colorectal cancer. The anticancer property of fucoidan is exerted at different stages of cancer formation and progression such as tumorigenesis and tumor cell proliferation, tumor progression or metastasis, angiogenesis, and apoptosis of cancer cells.

Fucoidan inhibits tumorigenesis and tumor cell proliferation by reducing the expression of cyclin D1, cyclin E and cyclin-dependent kinases (CDK2, CDK4). Cyclin D1 is the important regulator of cell cycle progression, and it forms the subunit for the cyclin-dependent kinases (Alao 2007). Overexpression of cyclin D1 is associated with the development of different types of cancers. Inhibiting the expression of cyclin D1 by fucoidan treatment can arrest the cell cycle and prevent tumor cell proliferation in colorectal cancer. The inhibition of mRNA expression of cyclin D1, CDK4 and CDK6 was shown to inhibit cell proliferation in Ht-29 human colon cancer cells (Yun et al., 2016). Concurrently, fucoidan also promotes the expression of CDK inhibitors such as p21WAF1/C1P1 and p27KIP1 (Park et al., 2017). The CDK inhibitors bind with CDK2 and CDK4 and arrest the cell cycle. In addition to the inhibition of cyclins and CDK, fucoidan also inhibits cell survival and proliferation by downregulation of the PI3K/Akt/mTOR signaling pathway (Han, Lee, and Lee 2015). Fucoidan downregulates the PI3K/Akt/mTOR pathway by inhibiting the phosphorylation of phosphoinositide 3-kinase (PI3K) and inhibiting the formation of its downstream effect Akt and mammalian target of rapamycin (mTOR) (Han, Lee, and Lee 2015). The response of colorectal cancer cells to fucoidan can be increased by silencing genes associated with the signaling of cell survival and cancer progression (Yun et al., 2016).

Fucoidan inhibits tumor progression by preventing migration or metastasis of malignant cells. The preventive action of fucoidan on metastasis of colon cancer cells involves decreasing the expression of matrix metalloproteinase-2 (MMP-2). Metastasis of malignant cells requires the degradation of extracellular matrix proteins by the action of MMP-2. Decreased expression of MMP-2 prevents metastasis and migration of cancer cells by keeping the extracellular matrix intact. Additionally, downregulation of the PI3K/Akt/mTOR signaling pathway also reduces the motility of colorectal cancer cells and induces apoptosis (Han, Lee, and Lee 2015). Fucoidan induces apoptosis by increasing the synthesis and activation of caspases such as caspase-8, caspase-9, caspase-7 and caspase-3 and degradation of poly(ADP-ribose) polymerase (Kim et al., 2010). Fucoidan destabilizes the mitochondrial membrane and increases its permeability which releases cytochrome c inducing the intrinsic pathway of apoptosis. Fucoidan also increases the levels of tumor necrosis factor-α and death receptor proteins involved in the extrinsic pathway of apoptosis.

Conclusion

Fucoidan polysaccharide from brown macroalgae is an excellent prebiotic that can help the beneficial microbes to establish dominance in the gastrointestinal tract while suppressing the colonization of potential pathogens. Fucoidan administration is also beneficial in regaining the microbial colonies in the gut following antibiotic treatment. Administration of fucoidan alone or in combination with other therapeutic agents is useful in alleviating inflammation in the gastrointestinal tract and in the treatment of gastrointestinal carcinoma. Elucidating the molecular mechanisms behind its cellular action would bridge the knowledge gap between the bioactive properties of fucoidan and developing it as a functional food ingredient or therapeutic agent for improving gastrointestinal function.

Disclaimer

None.

References

Alao, John P. 2007. "The Regulation of Cyclin D1 Degradation: Roles in Cancer Development and the Potential for Therapeutic Invention." *Molecular Cancer* 6: 1–16. https://doi.org/10.1186/1476-4598-6-24.

Cotas, João, Diana Pacheco, Ana M M Gonçalves, Paulo Silva, Loïc G Carvalho, and Leonel Pereira. 2021. "Seaweeds' Nutraceutical and Biomedical Potential in Cancer Therapy : A Concise Review." *Journal of Cancer Metastasis and Treatment* 7 (13). https://doi.org/10.20517/2394-4722.2020.134.

Cui, Wentong, Yuanyuan Zheng, Quanbin Zhang, Jing Wang, Limin Wang, Wenzhe Yang, Chenyang Guo, Weidong Gao, Xiaomin Wang, and Dali Luo. 2014. "Low-Molecular-Weight Fucoidan Protects Endothelial Function and Ameliorates Basal Hypertension in Diabetic Goto-Kakizaki Rats." *Laboratory Investigation* 94 (4): 382–93. https://doi.org/10.1038/labinvest.2014.12.

Fei, Na, and Liping Zhao. 2013. "An Opportunistic Pathogen Isolated from the Gut of an Obese Human Causes Obesity in Germfree Mice." *ISME Journal* 7 (4): 880–84. https://doi.org/10.1038/ismej.2012.153.

Flórez-Méndez, Jennyfer, and Leticia González. 2019. "Role of the Consumption of Fucoidans and Beta-Glucans on Human Health: An Update of the Literature." *Revista Chilena de Nutricion* 46 (6): 768–75. https://doi.org/10.4067/S0717-75182019000600768.

Han, Yong Seok, Jun Hee Lee, and Sang Hun Lee. 2015. "Fucoidan Inhibits the Migration and Proliferation of HT-29 Human Colon Cancer Cells via the Phosphoinositide-3 Kinase/Akt/Mechanistic Target of Rapamycin Pathways." *Molecular Medicine Reports* 12 (3): 3446–52. https://doi.org/10.3892/mmr.2015.3804.

Hrncir, Tomas. 2022. "Gut Microbiota Dysbiosis: Triggers, Consequences, Diagnostic and Therapeutic Options." *Microorganisms* 10 (3). https://doi.org/10.3390/microorganisms10030578.

Hwang, Pai An, Chien S. Y., Chan Y. L., Lu M. K., Wu C. H., Kong Z. L., and Chang-Jer Wu. Inhibition of Lipopolysaccharide (LPS)-induced inflammatory responses by *Sargassum hemiphyllum* sulfated polysaccharide extract in RAW 264.7 macrophage cells. *J Agric Food Chem* 2011; 59(5): 20628.

Hwang, Pai An, Nam Nhut Phan, Wen Jung Lu, Bui Thi Ngoc Hieu, and Yen Chang Lin. 2016. "Low-Molecular-Weight Fucoidan and High-Stability Fucoxanthin from Brown Seaweed Exert Prebiotics and Anti-Inflammatory Activities in Caco-2 Cells." *Food and Nutrition Research* 60: 1–9. https://doi.org/10.3402/fnr.v60.32033.

Jeong, Yong-Tae, Yong Deuk Kim, Young-Mi Jung, Dong-Chan Park, Dong-Sub Lee, Sae-Kwang Ku, Xian Li, Yue Lu, Guang Hsuan Chao, Keuk-Jun Kim, Jai-Youl Lee, Moon-Chang Baek, Wonku Kang, Seung-Lark Hwang, Hyeun Wook Chang. 2013. "Low Molecular Weight Fucoidan Improves Endoplasmic Reticulum Stress-Reduced Insulin Sensitivity through AMP-Activated Protein Kinase Activation in L6 Myotubes and Restores Lipid Homeostasis in a Mouse Model of Type 2 Diabetes." *Molecular Pharmacology* 84 (1): 147 LP – 157. https://doi.org/10.1124/mol.113.085100.

Kim, Eun J., So Y. Park, Jae Yong Lee, and Jung H. Y. Park. 2010. "Fucoidan Present in Brown Algae Induces Apoptosis of Human Colon Cancer Cells." *BMC Gastroenterology* 10. https://doi.org/10.1186/1471-230X-10-96.

Kim, Kui-Jin, Kye-Yoon Yoon, and Boo-Yong Lee. 2012. "Fucoidan Regulate Blood Glucose Homeostasis in C57BL/KSJ M+/+db and C57BL/KSJ Db/Db Mice." *Fitoterapia* 83 (6): 1105–9. https://doi.org/https://doi.org/10.1016/j.fitote.2012.04.027.

Lee, A. S., Gibson, D. L., Zhang, Y., Sham, H. P., Vallance, B. A., & Dutz, J. P. (2010). Gut barrier disruption by an enteric bacterial pathogen accelerates insulitis in NOD mice. *Diabetologia,* 53(4), 741–748. https://doi.org/10.1007/s00125-009-1626-y.

Liyanage, Nilanga, D. P. Nagahawatta, and Thilina U. Jayawardena. 2023. *"The Role of Seaweed Polysaccharides in Gastrointestinal Health: Protective Effect against Inflammatory Bowel Disease."*

Luthuli, Sibusiso, Wu. Siya, Yang Cheng, Xiaoli Zheng, Mingjiang Wu, and Haibin Tong. 2019. "Therapeutic Effects of Fucoidan : A Review on Recent Studies." *Marine Drugs* 17 (487): 1–15.

Ma, Zhihui, Zhiyou Yang, Xinyue Feng, Jiahang Deng, Chuantong He, Rui Li, Yuntao Zhao, Yuewei Ge, Yongping Zhang, Cai Song, Saiyi Zhong. 2022. "The Emerging Evidence for a Protective Role of Fucoidan from *Laminaria japonica* in Chronic Kidney Disease-Triggered Cognitive Dysfunction." *Marine Drugs* 20: 258. https://doi.org/https://doi.org/103390/md20040258.

Mukhamejanov, Emil, Galina Kon, Sara Erjanova, Aray Kirgizbaeva, and Elizabeth Muhamadieva. 2019. "Fucoidan - New Principle Prevention and Treatment of Diabetes." *Journal of Pharmacy and Pharmacology* 7 (6): 316–22. https://doi.org/10.17265/2328-2150/2019.06.005.

Niki, Toshiro, Shoko Tsutsui, Shigeru Hirose, Sachiko Aradono, Yasushi Sugimoto, Keisuke Takeshita, Nozomu Nishi, and Mitsuomi Hirashima. 2009. "Galectin-9 Is a

High Affinity IgE-Binding Lectin with Anti-Allergic Effect by Blocking IgE-Antigen Complex Formation." *Journal of Biological Chemistry* 284 (47): 32344–52. https://doi.org/10.1074/jbc.M109.035196.

Park, Hye Young, Shin Hyung Park, Jin Woo Jeong, Dahye Yoon, Min Ho Han, Dae Sung Lee, Grace Choi, Mi-Jin Yim, Jeong Min Lee, Do-Hyung Kim, Gi-Young Kim, Il-Whan Choi, Suhkmann Kim, Heui-Soo Kim, Hee-Jae Cha, Yung Hyun Choi. 2017. "Induction of P53-Independent Apoptosis and G1 Cell Cycle Arrest by Fucoidan in HCT116 Human Colorectal Carcinoma Cells." *Marine Drugs* 15 (6): 1–14. https://doi.org/10.3390/md15060154.

Rajapakse, Niranjan, and Se Kwon Kim. 2011. *Nutritional and Digestive Health Benefits of Seaweed. Advances in Food and Nutrition Research*. 1st ed. Vol. 64. Elsevier Inc. https://doi.org/10.1016/B978-0-12-387669-0.00002-8.

Ricós-Muñoz, Neus, Sergi Maicas, and María Consuelo Pina-Pérez. 2021. "*Probiotic Lactobacillus Reuteri Growth Improved under Fucoidan Exposure,*" 106. https://doi.org/10.3390/foods_2020-07724.

Shan, Xindi, Xin Liu, Jiejie Hao, Chao Cai, Fei Fan, Yunlou Dun, Xiaoliang Zhao, Xiaoxiao Liu, Chunxia Li, and Guangli Yu. 2016. "In Vitro and in Vivo Hypoglycemic Effects of Brown Algal Fucoidans." *International Journal of Biological Macromolecules* 82: 249–55. https://doi.org/https://doi.org/10.1016/j.ijbiomac.2015.11.036.

Shang, Qingsen, Guanrui Song, Meifang Zhang, Jingjing Shi, Cuiying Xu, Jiejie Hao, Guoyun Li, and Guangli Yu. 2017. "Dietary Fucoidan Improves Metabolic Syndrome in Association with Increased Akkermansia Population in the Gut Microbiota of High-Fat Diet-Fed Mice." *Journal of Functional Foods* 28: 138–46. https://doi.org/https://doi.org/10.1016/j.jff.2016.11.002.

Shang, Qingsen, Xindi Shan, Chao Cai, Jiejie Hao, Guoyun Li, and Guangli Yu. 2016. "Dietary Fucoidan Modulates the Gut Microbiota in Mice by Increasing the Abundance of Lactobacillus and Ruminococcaceae." *Food & Function* 7 (7): 3224–32. https://doi.org/10.1039/c6fo00309e.

Sun, Ting, Hui Liang, Meilan Xue, Ying Liu, Anjing Gong, Yushan Jiang, Yimin Qin, Jia Yang, and Danyang Meng. 2020. "Protective Effect and Mechanism of Fucoidan on Intestinal Mucosal Barrier Function in NOD Mice." *Food and Agricultural Immunology* 31 (1): 922–36. https://doi.org/10.1080/09540105.2020.1789071.

Tanino, Yuka, Takashi Hashimoto, Takao Ojima, and Masashi Mizuno. 2016. "F-Fucoidan from *Saccharina japonica* Is a Novel Inducer of Galectin-9 and Exhibits Anti-Allergic Activity." *Journal of Clinical Biochemistry and Nutrition* 59 (1): 25–30. https://doi.org/10.3164/jcbn.15-144.

Venardou, Brigkita, John V. O'Doherty, Shane Maher, Marion T. Ryan, Vivian Gath, Rajeev Ravindran, Claire Kiely, Gaurav Rajauria, Marco Garcia-Vaquero, and Torres Sweeney. 2022. "Potential of a Fucoidan-Rich *Ascophyllum nodosum* Extract to Reduce Salmonella Shedding and Improve Gastrointestinal Health in Weaned Pigs Naturally Infected with Salmonella." *Journal of Animal Science and Biotechnology* 13 (1): 1–16. https://doi.org/10.1186/s40104-022-00685-4.

Wang, Jing, Huaide Liu, Ning Li, Quanbin Zhang, and Hong Zhang. 2014. "The Protective Effect of Fucoidan in Rats with Streptozotocin-Induced Diabetic Nephropathy." *Marine Drugs* 12 (6): 3292–3306. https://doi.org/10.3390/md12063292.

Wang, Lilong, Chunqing Ai, Chengrong Wen, Yimin Qin, Zhengqi Liu, Linlin Wang, Yue Gong, Changyu Su, Zhongfu Wang, and Shuang Song. 2020. "Fucoidan Isolated from *Ascophyllum nodosum* Alleviates Gut Microbiota Dysbiosis and Colonic Inflammation in Antibiotic-Treated Mice." *Food & Function* 11 (6): 5595–5606. https://doi.org/10.1039/d0fo00668h.

Yang, Jin Young, and Sun Young Lim. 2021. "Fucoidans and Bowel Health." *Marine Drugs* 19 (8): 1–13. https://doi.org/10.3390/md19080436.

Yang, Wenzhe, Xinfeng Yu, Quanbin Zhang, Qingjun Lu, Jing Wang, Wentong Cui, Yuanyuan Zheng, Xiaomin Wang, and Dali Luo. 2013. "Attenuation of Streptozotocin-Induced Diabetic Retinopathy with Low Molecular Weight Fucoidan via Inhibition of Vascular Endothelial Growth Factor." *Experimental Eye Research* 115: 96–105. https://doi.org/https://doi.org/10.1016/j.exer.2013.06.011.

Yuan, Jingyi, Song Qin, Wenjun Li, Yubing Zhang, Yuting Wang, Xiulian Chang, and Lili Li. 2021. "Microbiome-Metabolomics Reveals Prebiotic Benefits of Fucoidan Supplementation in Mice." *Journal of Marine Science and Engineering* 9 (5). https://doi.org/10.3390/jmse9050505.

Yun, Chul Won, Seungpil Yun, Jun Hee Lee, Yong Seok Han, Yeo Min Yoon, Daniel An, and Sang Hun Lee. 2016. "Silencing Prion Protein in Ht29 Human Colorectal Cancer Cells Enhances Anticancer Response to Fucoidan." *Anticancer Research* 36 (9): 4449–58. https://doi.org/10.21873/anticanres.10989.

Zhou, Kequan. 2017. "Strategies to Promote Abundance of *Akkermansia muciniphila*, an Emerging Probiotics in the Gut, Evidence from Dietary Intervention Studies." *Journal of Functional Foods* 33 (2004): 194–201. https://doi.org/10.1016/j.jff.2017.03.045.

Chapter 7

The Roles of Actin Cytoskeleton in Hepatic Stellate Cell Activation and Liver Fibrosis

Archittapon Nokkeaw[1,2,3]
Pannathon Thamjamrassri[1,2,3]
Pisit Tangkijvanich[1,2,*]
and Chaiyaboot Ariyachet[1,2,†]

[1]Department of Biochemistry, Faculty of Medicine, Chulalongkorn University, Bangkok, Thailand
[2]Center of Excellence in Hepatitis and Liver Cancer, Faculty of Medicine, Chulalongkorn University, Bangkok, Thailand
[3]Medical Biochemistry Program, Department of Biochemistry, Faculty of Medicine, Chulalongkorn University, Bangkok, Thailand

Abstract

Hepatic stellate cells (HSCs) are the key contributor in excessive accumulation of extracellular matrix (ECM) causing liver fibrosis which occurs in virtually all chronic liver diseases. In normal liver, HSCs remain dormant and store retinol in lipid droplets. Yet after injuries, quiescent HSCs become activated and transform into ECM-producing myofibroblasts with increased contractility due to high expression of alpha smooth muscle actin (α-SMA). Once thought merely as a key marker for activated HSCs, α-SMA has played a major control in mechanotransduction that promotes fibrogenic signaling in HSCs. In this review chapter, we will discuss functions of α-SMA in HSC activation and its potential as a therapeutic target for antifibrotic therapy.

* Corresponding Author's Email: pisittkvn@yahoo.com.
† Corresponding Author's Email: Chaiyaboot.A@chula.ac.th.

In: Advances in Biology. Volume 4
Editor: Charles D. Grant
ISBN: 979-8-88697-999-2

Keywords: cytoskeleton, smooth muscle actin, extracellular matrix, hepatic stellate cell, liver fibrosis, mechanotransduction

Introduction

Liver fibrosis is an abnormal wound-healing process of the liver characterized by the accumulation of excessive extracellular matrix (ECM) caused by hepatitis, alcohol, or non-alcoholic steatohepatitis (NASH) (Pellicoro et al. 2014; Stalnikowitz and Weissbrod 2003). Advanced liver fibrosis can lead to the development of cirrhosis and eventually hepatocellular carcinoma (Bataller and Brenner 2005), both of which cause many deaths worldwide (Huang et al. 2023; Rumgay et al. 2022). Hepatic stellate cells (HSCs), which undergo a transition from a quiescent to an active state in response to hepatic injuries, are primarily responsible for the development of liver fibrosis (Hou and Syn 2018). This transition results in excessive production of ECM, especially type I and type III collagen, and the formation of a fibrous scar (Yan et al. 2021; Tsuchida and Friedman 2017).

Cytoskeleton is a network of filaments found inside the cytoplasm of a cell that is responsible for a number of different cellular physiological processes, such as the formation of stress fibers, adhesion, migration, apoptosis, and intercellular communication (Wickstead and Gull 2011; Cui et al. 2014). Actin, microtubules, and intermediate filaments are the three fundamental types of cytoskeletal filaments that facilitate the aforementioned functions. Among these cytoskeletons, actin is considered the most dynamic cytoskeleton (Hohmann and Dehghani 2019). Actin dynamics are crucial in the regulation of various cellular processes (Villalonga and Ridley 2013). During HSC activation, alteration of the actin cytoskeleton was observed, indicating its important roles in HSC activation (Yee 1998). Indeed, knockout expression of actin leads to decreased liver fibrosis and ECM production in rodent models (D. C. Rockey et al. 2019). This article focuses on the latest discoveries about the actin cytoskeleton dynamics in HSC activation and hepatic fibrosis.

HSCs and Matrix Stiffness

HSCs are non-parenchymal cells that account for around 10% of the resident cells in a normal liver (Geerts 2001). HSCs, pericyte cells of the liver, reside in the space of Disse between the fenestrated liver endothelium and hepatocytes (Hellerbrand 2013). HSCs, similar to pericytes in other organs, control blood flow and produce ECM under inflammatory conditions. Furthermore, HSCs are involved in vitamin A storage, support of liver regeneration, and immunoregulation (Fujita and Narumiya 2016; Hellerbrand 2013). HSCs are in a quiescent state in normal conditions, but they became activated following the liver injury (Tacke and Weiskirchen 2012), which activates fibrotic signaling pathways such as TGF-β (Dewidar et al. 2019), JAK/STAT (Kagan et al. 2017), PI3K/Akt (Cai et al. 2017), Wnt/β-catenin (Ge et al. 2014), Hedgehog (Sicklick et al. 2005), and Hippo (Mannaerts et al. 2015) signaling pathways. In the activated state, HSCs transform into highly proliferative, myofibroblast-like cell types that produce and secrete excess amounts of ECM, and its accumulation is amplified by the production of TIMPs, an inhibitor of ECM-degrading enzymes (Tacke and Weiskirchen 2012; Mormone, George, and Nieto 2011). As a result, ECM deposition was significantly increased, enhancing matrix stiffness (Ortiz et al. 2021; Herrera, Henke, and Bitterman 2018).

Matrix stiffness is involved with pro-fibrotic signaling pathways, including TGF-β (Wells and Discher 2008), JAK/STAT (Hu et al. 2023), PI3K/Akt (Gao et al. 2020; Liu et al. 2021), Wnt/β-catenin (Du et al. 2016), Hedgehog (Yu et al. 2020), and Hippo (Chang et al. 2020; Fu et al. 2022), indicating a possible association between matrix stiffness and HSC activation. Many studies have shown that when HSCs are grown on soft supports, they are unable to undergo transdifferentiation, but when cultured on stiff supports, they transform into myofibroblast-like cells (Olsen et al. 2011; Kostallari et al. 2022; Lachowski et al. 2019). The stiff matrix also downregulates ECM-degrading enzymes called matrix metalloproteinases (MMPs) and promotes the secretion of their inhibitors (TIMPs), which further intensifies the stiffness (Lachowski et al. 2019). These findings point to the importance of the cellular microenvironment in controlling the phenotypic change in HSCs.

Actin and Fibrosis

Actin, the most common protein that plays a critical role in cellular processes such as cell division, migration, and signaling in eukaryotic cells, expresses three main isoforms in vertebrates, including the α-, β-, and γ-isoforms. Each isoform differs by a few amino acids toward the N-terminus. There are two forms of actin: G-actin (also known as globular actin) and F-actin (also known as fibrous actin) (Dominguez and Holmes 2011). The functional role of actin in cellular processes requires the rapid organization of actin between monomers and filaments. This dynamic process is regulated by numerous actin-binding proteins (ABPs) and their upstream regulators (Hansen and Kwiatkowski 2013; Rajakylä and Vartiainen 2014), which have been reported to be involved in HSC activation (Kato et al. 1999; Rombouts et al. 2012; Hironaka et al. 2023).

Following the liver injury, HSCs became activated and extensively expressed alpha smooth muscle actin (α-SMA), a key biomarker of HSC activation (Carpino et al. 2005). Conversely, quiescent HSCs appear to only express β- and γ-actin isoforms and are identified by the lack or disorganization of actin stress fibers (D. C. Rockey et al. 2019; Shi, Ren, and Rockey 2020). α-SMA is associated with the cellular motility and contractility of HSCs as knockdown expression of *Acta2*, which encodes α-SMA, results in reduced cell motility and contractility (Don C. Rockey, Weymouth, and Shi 2013). Scar contractures and fibrosis develop when there is an excessive amount of wound contraction, indicating an α-SMA role in fibrosis development (Ishise et al. 2015). Furthermore, scar contractures can trigger many fibrogenic pathways, thus creating a positive feedback loop that additionally aggravates fibrosis (Kuehlmann et al. 2020). The depletion of α-SMA in HSCs does, in fact, result in a reduction in liver fibrosis (D. C. Rockey et al. 2019). Likewise, overexpression of microRNA-223, a negative regulator of *Acta2*, causes suppression of activation phenotypes, including reductions in collagen production, proliferation, and cellular contractility (Ariyachet et al. 2022).

Mechanisms of Actin Involved in Liver Fibrosis

As previously mentioned, actin plays a significant role in the development of liver fibrosis; therefore, it is essential to comprehend the mechanism by which actin regulates liver fibrosis in order to better understand the pathology of the disease, which will assist in the development of more effective treatments for liver fibrosis.

Mechanotransduction, the transformation of mechanical forces into biochemical signals, has been suggested to play a crucial role in the scarring process due to its strong connection with fibrosis (Duscher et al. 2014). In this process, many receptors play a key role as mediators, allowing signal transduction from the biomechanical environment. These receptors include stretch-activated ion channels, growth factor receptors, G protein-coupled receptors, and integrins (Duscher et al. 2014). After these receptors receive the biomechanical force, they transfer the signal through many signaling pathways that are associated with fibrogenesis by promoting α-SMA expression, including FAK (Hetzler et al. 2019), Wnt (Corbett, Mann, and Mann 2015), Hippo (Li et al. 2018), and TGF-β (Zhang et al. 2021) signaling pathways. Moreover, these pathways can also activate many signaling molecules, especially Rho GTPases (Hoon, Tan, and Koh 2016; Habas, Kato, and He 2001; Ebrahimi et al. 2022; Jang, Kim, and Bae 2020; Shimada, Staten, and Rajagopalan 2011), and consequently activate Rho kinases, which are major regulators of the actin cytoskeleton and promote the incorporation of α-SMA into the actin cytoskeleton (Amano, Nakayama, and Kaibuchi 2010; Vardouli et al. 2008). α-SMA incorporated into focal adhesions, mechanosensors, and stress fibers can enhance ECM tractional remodeling, ultimately resulting in scar formation, which contributes to fibrosis (J. Wang, Zohar, and McCulloch 2006; Cox and Erler 2011). In this regard, α-SMA filaments provide linkage between mechanosensory elements, which receive biomechanical signals, and their targets necessary for mechanotransduction (Wen et al. 2022; J. Wang, Zohar, and McCulloch 2006). The incorporation of α-SMA into stress fibers can increase myofibroblast contraction and tensional forces on nearby ECM, increasing ECM stiffness and further promoting α-SMA expression (Adapala et al. 2021; Liu et al. 2021). Interestingly, mechanical force can influence the conformation of the ECM, which can expose hidden domains or binding sites of the ECM, alter spatial density, and release stored factors (Duscher et al. 2014). ECM can act as a reservoir for signaling molecules, mostly through electrostatic interaction between the negative charge of proteoglycans and the positive charge of the signaling molecule (Franz et al. 2011). TGF-β is stored in the ECM in a latent protein complex called latent TGF-β binding protein, or LTBP, which is covalently cross-linked to ECMs (Hayashi and Sakai 2012). The possible mechanism for the release of TGF-β from the latent complex is integrin-mediated traction, which changes the conformation of the latent complex (Wipff and Hinz 2008). These effects can lead to the development of fibrosis, especially the release of stored TGF-β from the ECM to drive HSC activation (Duscher et al. 2014).

A study shows that matrix stiffness promotes activation of HSCs by the CD36-AKT axis via mechanosignaling (Liu et al. 2021). In a stiff environment, CD36 induces AKT phosphorylation. AKT is associated with the fibrogenic signaling pathway, and the inhibition of AKT signaling significantly reduces α-SMA levels (R. Wang et al. 2019; Pei et al. 2018).

Furthermore, Nesprin is suggested to be involved in fibrogenesis via the actin cytoskeleton. Nuclear envelope spectrin-repeat protein (Nesprin) is a scaffold protein that connects nuclei to the actin cytoskeleton. They contain two domains: the Claponin homology domain and the KASH domain. The Calponin homology is located at the N-terminus and interacts with F-actin, while the C-terminal KASH domain extends into the perimuclear space and interacts with anchored proteins of the inner nuclear membrane (Rajgor and Shanahan 2013; Xu et al. 2022). Nesprin is part of the linker of the nucleoskeleton and cytoskeleton (LINC) complex. These complexes play a major role in the transduction of mechanical signals to the nucleus via cytoskeletal linkage (Cho, Irianto, and Discher 2017). The strong tension force applied to the nuclear envelope can cause nuclear deformation, which in turn activates many downstream mechanisms associated with gene expression regulation. The disruption of the LINC complex by interfering with Nesprin results in a quiescent phenotype of cells despite being cultured on a stiff matrix (Guixé-Muntet et al. 2020). Interestingly, actin polymerization can interrupt the interaction between LEMD3 and SMAD2/3. LEMD3 antagonizes the TGF-β-SMAD2/3 fibrogenic signaling pathway; thus, actin polymerization can promote fibrogenesis (Chambers et al. 2018). Additionally, SMAD3 can induce the expression of α-SMA and increase its incorporation into the stress fibers (Vardouli et al. 2008), thus further promoting fibrogenesis via a positive feedback loop. Importantly, actin remodeling is considered to play a key role in the Hippo signaling pathway. Stress fibers and RHO GTPases activation can promote YAP/TAZ nuclear localization, which in turn promotes transcription of fibrogenic genes (Seo and Kim 2018). Lastly, matrix stiffness influences cytoskeleton arrangement and triggers MRTF-A nuclear translocation by freeing MRTF-A from the G-actin/MRTF-A complex, which consequently induces expression of α-SMA and other fibrotic genes (Shi, Ren, and Rockey 2020).

Therapeutic Potential of Targeting α-SMA

Given the crucial roles of α-SMA in HSC activation and liver fibrosis, this gene has emerged as a potential therapeutic target for antifibrotic therapy. Indeed, inhibition of α-SMA function can impair fibrotic signaling pathways and prevent activation of HSCs. For example, blocking peptides that decrease the traction force generated by α-SMA can inhibit the nuclear localization of YAP/TAZ, two transcriptional co-activators that regulate gene expression in response to mechanical cues. This suggests that the contractility of cells from α-SMA during HSC transdifferentiation is critical for YAP/TAZ-mediated mechanotransduction (Caliari et al. 2016). In addition, prostaglandin E_2 (PGE_2), a naturally occurring lipid mediator, has been shown to inhibit α-SMA transcription during myofibroblast differentiation. The downregulation of α-SMA expression by PGE_2 is mediated by the activation of prostaglandin E receptor 2 (EP2), which stimulates the cAMP-PKA pathway and consequently inhibits p38 MAPK phosphorylation. p38 is a subgroup of the MAPKs that are exclusively activated by environmental stresses and inflammatory signals (Canovas and Nebreda 2021). As phosphorylated p38 could promote SRF transcription, PGE_2 reduces the nuclear accumulation of MRTF-A/SRF complexes, thus blocking transcription of α-SMA (Penke et al. 2014). Sesamol (SML) could alleviate liver fibrosis by downregulating lysophosphatidic acid receptor (LPAR)-1 and -3, which are able to suppress α-SMA gene and protein expression (Elrazik, El-Mesery, and El-Shishtawy 2022). Lysophosphatidic acid (LPA) is a phospholipid that regulates various cellular processes by binding to LPARs and triggering a signaling cascade, which is involved in liver fibrosis by promoting α-SMA expression (Geraldo et al. 2021; Kaffe, Magkrioti, and Aidinis 2019). Farnesoid X receptor (FXR) is a nuclear receptor that plays a critical role in regulating bile acid homeostasis and lipid metabolism (Claudel, Staels, and Kuipers 2005). Additionally, FXR appears to inhibit HSC activation via increasing PPARγ expression, which negatively regulates expression of α-SMA (Renga et al. 2011; Stefano et al. 2005).

Overall, targeting α-SMA represents a promising approach for the development of antifibrotic therapies for the treatment of liver fibrosis. However, further studies are required to identify molecules or methods that can specifically downregulate α-SMA in HSCs for antifibrotic therapy and to determine the safety and efficacy of these approaches in human relevant models.

Conclusion

In summary, actin is considered the most dynamic cytoskeleton and is crucial in the regulation of various cellular processes. Alterations in the actin cytoskeleton were observed during HSC activation, indicating its important roles in HSC activation and liver fibrosis. Many studies have discovered that mechanotransduction associated with the actin cytoskeleton is a key player in fibrogenesis. The actin cytoskeleton links mechanosensory elements with their targets to appropriately transduce the biomechanical to biochemical signals that are involved in fibrosis development. The possible mechanisms behind this function may involve Nesprin, LEMD3, MRTF-A, and the Hippo signaling pathway. Incorporation of α-SMA into stress fibers can enhance myofibroblast contraction, thus increasing tensional force on the ECM. The mechanical force from ECM can stimulate mechanosensory receptors (i.e., CD36, integrin, etc.), which trigger many downstream signaling cascades, including Rho GTPases. Rho GTPases regulate actin remodeling, a process that is associated with mechanotransduction involving Nesprin, LEMD3, MRTF-A, and the YAP/TAZ in the Hippo signaling pathway. Although knowledge around the actin cytoskeleton and HSC activation has advanced recently, there are still areas to be explored. Understanding the processes of mechanotransduction might assist in discovering effective novel therapies for the treatment of liver fibrosis.

Disclaimer

None.

References

Adapala, R. K., V. Katari, L. R. Teegala, S. Thodeti, S. Paruchuri, and C. K. Thodeti. 2021. "TRPV4 Mechanotransduction in Fibrosis." *Cells* 10 (11).

Amano, M., M. Nakayama, and K. Kaibuchi. 2010. "Rho-kinase/ROCK: A key regulator of the cytoskeleton and cell polarity." *Cytoskeleton (Hoboken)* 67 (9): 545-54.

Ariyachet, C., N. Chuaypen, P. Kaewsapsak, N. Chantaravisoot, D. Jindatip, S. Potikanond, and P. Tangkijvanich. 2022. "MicroRNA-223 Suppresses Human Hepatic Stellate Cell Activation Partly via Regulating the Actin Cytoskeleton and Alleviates Fibrosis in Organoid Models of Liver Injury." *Int J Mol Sci* 23 (16).

Bataller, R., and D. A. Brenner. 2005. "Liver fibrosis." *J Clin Invest* 115 (2): 209-18.

Cai, C. X., H. Buddha, S. Castelino-Prabhu, Z. Zhang, R. S. Britton, B. R. Bacon, and B. A. Neuschwander-Tetri. 2017. "Activation of Insulin-PI3K/Akt-p70S6K Pathway in Hepatic Stellate Cells Contributes to Fibrosis in Nonalcoholic Steatohepatitis." *Dig Dis Sci* 62 (4): 968-978.

Caliari, S. R., M. Perepelyuk, B. D. Cosgrove, S. J. Tsai, G. Y. Lee, R. L. Mauck, R. G. Wells, and J. A. Burdick. 2016. "Stiffening hydrogels for investigating the dynamics of hepatic stellate cell mechanotransduction during myofibroblast activation." *Sci Rep* 6: 21387.

Canovas, Begoña, and Angel R. Nebreda. 2021. "Diversity and versatility of p38 kinase signalling in health and disease." *Nature Reviews Molecular Cell Biology* 22 (5): 346-366.

Carpino, G., S. Morini, S. Ginanni Corradini, A. Franchitto, M. Merli, M. Siciliano, F. Gentili, A. Onetti Muda, P. Berloco, M. Rossi, A. F. Attili, and E. Gaudio. 2005. "Alpha-SMA expression in hepatic stellate cells and quantitative analysis of hepatic fibrosis in cirrhosis and in recurrent chronic hepatitis after liver transplantation." *Dig Liver Dis* 37 (5): 349-56.

Chambers, D. M., L. Moretti, J. J. Zhang, S. W. Cooper, D. M. Chambers, P. J. Santangelo, and T. H. Barker. 2018. "LEM domain-containing protein 3 antagonizes TGFβ-SMAD2/3 signaling in a stiffness-dependent manner in both the nucleus and cytosol." *J Biol Chem* 293 (41): 15867-15886.

Chang, Yu-Chiuan, Jhen-Wei Wu, Chueh-Wen Wang, and Anna C.-C. Jang. 2020. "Hippo Signaling-Mediated Mechanotransduction in Cell Movement and Cancer Metastasis." *Frontiers in Molecular Biosciences* 6.

Cho, S., J. Irianto, and D. E. Discher. 2017. "Mechanosensing by the nucleus: From pathways to scaling relationships." *J Cell Biol* 216 (2): 305-315.

Claudel, Thierry, Bart Staels, and Folkert Kuipers. 2005. "The Farnesoid X Receptor." *Arteriosclerosis, Thrombosis, and Vascular Biology* 25 (10): 2020-2030.

Corbett, Laura, Jelena Mann, and Derek A. Mann. 2015. "Non-Canonical Wnt Predominates in Activated Rat Hepatic Stellate Cells, Influencing HSC Survival and Paracrine Stimulation of Kupffer Cells." *PLOS ONE* 10 (11): e0142794.

Cox, T. R., and J. T. Erler. 2011. "Remodeling and homeostasis of the extracellular matrix: implications for fibrotic diseases and cancer." *Dis Model Mech* 4 (2): 165-78.

Cui, X., X. Zhang, Q. Yin, A. Meng, S. Su, X. Jing, H. Li, X. Guan, X. Li, S. Liu, and M. Cheng. 2014. "F-actin cytoskeleton reorganization is associated with hepatic stellate cell activation." *Mol Med Rep* 9 (5): 1641-7.

Dewidar, B., C. Meyer, S. Dooley, and A. N. Meindl-Beinker. 2019. "TGF-β in Hepatic Stellate Cell Activation and Liver Fibrogenesis Updated 2019." *Cells* 8 (11).

Dominguez, R., and K. C. Holmes. 2011. "Actin structure and function." *Annu Rev Biophys* 40: 169-86.

Du, J., Y. Zu, J. Li, S. Du, Y. Xu, L. Zhang, L. Jiang, Z. Wang, S. Chien, and C. Yang. 2016. "Extracellular matrix stiffness dictates Wnt expression through integrin pathway." *Sci Rep* 6: 20395.

Duscher, D., Z. N. Maan, V. W. Wong, R. C. Rennert, M. Januszyk, M. Rodrigues, M. Hu, A. J. Whitmore, A. J. Whittam, M. T. Longaker, and G. C. Gurtner. 2014. "Mechanotransduction and fibrosis." *J Biomech* 47 (9): 1997-2005.

Ebrahimi, Nasim, Kamyar Kharazmi, Masoud Ghanaatian, Seyed Ali Miraghel, Yasaman Amiri, Seyede Sanaz Seyedebrahimi, Halimeh Mobarak, Elnaz Yazdani, Sahar Parkhideh, Michael R. Hamblin, and Amir Reza Aref. 2022. "Role of the Wnt and GTPase pathways in breast cancer tumorigenesis and treatment." *Cytokine & Growth Factor Reviews* 67: 11-24.

Elrazik, Nesma A. Abd, Mohamed El-Mesery, and Mamdouh M. El-Shishtawy. 2022. "Sesamol protects against liver fibrosis induced in rats by modulating lysophosphatidic acid receptor expression and TGF-β/Smad3 signaling pathway." *Naunyn-Schmiedeberg's Archives of Pharmacology* 395 (8): 1003-1016.

Franz, S., S. Rammelt, D. Scharnweber, and J. C. Simon. 2011. "Immune responses to implants - a review of the implications for the design of immunomodulatory biomaterials." *Biomaterials* 32 (28): 6692-709.

Fu, Minyang, Yuan Hu, Tianxia Lan, Kun-Liang Guan, Ting Luo, and Min Luo. 2022. "The Hippo signalling pathway and its implications in human health and diseases." *Signal Transduction and Targeted Therapy* 7 (1): 376.

Fujita, Tomoko, and Shuh Narumiya. 2016. "Roles of hepatic stellate cells in liver inflammation: a new perspective." *Inflammation and Regeneration* 36 (1): 1.

Gao, Xiangyu, Xiaona Qiao, Xiaoxia Xing, Jinya Huang, Jiali Qian, Yi Wang, Yawen Zhang, Xi Zhang, Miao Li, Jiefeng Cui, and Yehong Yang. 2020. "Matrix Stiffness-Upregulated MicroRNA-17-5p Attenuates the Intervention Effects of Metformin on HCC Invasion and Metastasis by Targeting the PTEN/PI3K/Akt Pathway." *Frontiers in Oncology* 10: 1563.

Ge, Wen-Song, Yao-Jun Wang, Jian-Xin Wu, Jian-Gao Fan, Ying-Wei Chen, and Liang Zhu. 2014. "β-catenin is overexpressed in hepatic fibrosis and blockage of Wnt/β-catenin signaling inhibits hepatic stellate cell activation." *Mol Med Rep* 9 (6): 2145-2151.

Geerts, Albert. 2001. "History, Heterogeneity, Developmental Biology, and Functions of Quiescent Hepatic Stellate Cells." *Semin Liver Dis* 21 (03): 311-336.

Geraldo, Luiz Henrique Medeiros, Tânia Cristina Leite de Sampaio Spohr, Rackele Ferreira do Amaral, Anna Carolina Carvalho da Fonseca, Celina Garcia, Fabio de Almeida Mendes, Catarina Freitas, Marcos Fabio dos Santos, and Flavia Regina Souza Lima. 2021. "Role of lysophosphatidic acid and its receptors in health and disease: novel therapeutic strategies." *Signal Transduction and Targeted Therapy* 6 (1): 45.

Guixé-Muntet, S., M. Ortega-Ribera, C. Wang, S. Selicean, I. Andreu, J. Z. Kechagia, C. Fondevila, P. Roca-Cusachs, J. F. Dufour, J. Bosch, A. Berzigotti, and J. Gracia-Sancho. 2020. "Nuclear deformation mediates liver cell mechanosensing in cirrhosis." *JHEP Rep* 2 (5): 100145.

Habas, R., Y. Kato, and X. He. 2001. "Wnt/Frizzled activation of Rho regulates vertebrate gastrulation and requires a novel Formin homology protein Daam1." *Cell* 107 (7): 843-54.

Hansen, Marc D. H., and Adam V. Kwiatkowski. 2013. "Chapter One - Control of Actin Dynamics by Allosteric Regulation of Actin Binding Proteins." In *International Review of Cell and Molecular Biology*, edited by Kwang W. Jeon, 1-25. Academic Press.

Hayashi, H., and T. Sakai. 2012. "Biological Significance of Local TGF-β Activation in Liver Diseases." *Front Physiol* 3: 12.

Hellerbrand, C. 2013. "Hepatic stellate cells--the pericytes in the liver." *Pflugers Arch* 465 (6): 775-8.

Herrera, J., C. A. Henke, and P. B. Bitterman. 2018. "Extracellular matrix as a driver of progressive fibrosis." *J Clin Invest* 128 (1): 45-53.

Hetzler, P. T., 3rd, B. C. Dash, S. Guo, and H. C. Hsia. 2019. "Targeting Fibrotic Signaling: A Review of Current Literature and Identification of Future Therapeutic Targets to Improve Wound Healing." *Ann Plast Surg* 83 (6): e92-e95.

Hironaka, Takanori, Noburo Takizawa, Yuto Yamauchi, Yuma Horii, and Michio Nakaya. 2023. "The well-developed actin cytoskeleton and Cthrc1 expression by actin-binding protein drebrin in myofibroblasts promote cardiac and hepatic fibrosis." *Journal of Biological Chemistry* 299 (3).

Hohmann, T., and F. Dehghani. 2019. "The Cytoskeleton-A Complex Interacting Meshwork." *Cells* 8 (4).

Hoon, J. L., M. H. Tan, and C. G. Koh. 2016. "The Regulation of Cellular Responses to Mechanical Cues by Rho GTPases." *Cells* 5 (2).

Hou, W., and W. K. Syn. 2018. "Role of Metabolism in Hepatic Stellate Cell Activation and Fibrogenesis." *Front Cell Dev Biol* 6: 150.

Hu, Q., Q. Bian, D. Rong, L. Wang, J. Song, H. S. Huang, J. Zeng, J. Mei, and P. Y. Wang. 2023. "JAK/STAT pathway: Extracellular signals, diseases, immunity, and therapeutic regimens." *Front Bioeng Biotechnol* 11: 1110765.

Huang, Daniel Q., Norah A. Terrault, Frank Tacke, Lise Lotte Gluud, Marco Arrese, Elisabetta Bugianesi, and Rohit Loomba. 2023. "Global epidemiology of cirrhosis — aetiology, trends and predictions." *Nature Reviews Gastroenterology & Hepatology*.

Ishise, Hisako, Barrett Larson, Yutaka Hirata, Toshihiro Fujiwara, Soh Nishimoto, Tateki Kubo, Ken Matsuda, Shigeyuki Kanazawa, Yohei Sotsuka, Kazutoshi Fujita, Masao Kakibuchi, and Kenichiro Kawai. 2015. "Hypertrophic scar contracture is mediated by the TRPC3 mechanical force transducer via NFkB activation." *Scientific Reports* 5 (1): 11620.

Jang, J. W., M. K. Kim, and S. C. Bae. 2020. "Reciprocal regulation of YAP/TAZ by the Hippo pathway and the Small GTPase pathway." *Small GTPases* 11 (4): 280-288.

Kaffe, E., C. Magkrioti, and V. Aidinis. 2019. "Deregulated Lysophosphatidic Acid Metabolism and Signaling in Liver Cancer." *Cancers (Basel)* 11 (11).

Kagan, Polina, Maya Sultan, Irina Tachlytski, Michal Safran, and Ziv Ben-Ari. 2017. "Both MAPK and STAT3 signal transduction pathways are necessary for IL-6-dependent hepatic stellate cells activation." *PLOS ONE* 12 (5): e0176173.

Kato, Masaki, Hiroaki Iwamoto, Nobuhiko Higashi, Rie Sugimoto, Koutaro Uchimura, Seiya Tada, Hironori Sakai, Makoto Nakamuta, and Hajime Nawata. 1999. "Role of Rho small GTP binding protein in the regulation of actin cytoskeleton in hepatic stellate cells." *Journal of Hepatology* 31 (1): 91-99.

Kostallari, E., B. Wei, D. Sicard, J. Li, S. A. Cooper, J. Gao, M. Dehankar, Y. Li, S. Cao, M. Yin, D. J. Tschumperlin, and V. H. Shah. 2022. "Stiffness is associated with hepatic stellate cell heterogeneity during liver fibrosis." *Am J Physiol Gastrointest Liver Physiol* 322 (2): G234-g246.

Kuehlmann, Britta, Clark A. Bonham, Isabel Zucal, Lukas Prantl, and Geoffrey C. Gurtner. 2020. Mechanotransduction in Wound Healing and Fibrosis. *Journal of Clinical Medicine* 9 (5).

Lachowski, Dariusz, Ernesto Cortes, Alistair Rice, David Pinato, Krista Rombouts, and Armando del Rio Hernandez. 2019. "Matrix stiffness modulates the activity of MMP-

9 and TIMP-1 in hepatic stellate cells to perpetuate fibrosis." *Scientific Reports* 9 (1): 7299.

Li, Li, Jie Zhou, Qiufang Li, Jianjun Xu, Jing Qi, and Hua Bian. 2018. "The inhibition of Hippo/Yap signaling pathway is required for magnesium isoglycyrrhizinate to ameliorate hepatic stellate cell inflammation and activation." *Biomedicine & Pharmacotherapy* 106: 83-91.

Liu, Zhikui, Huanye Mo, Runkun Liu, Yongshen Niu, Tianxiang Chen, Qiuran Xu, Kangsheng Tu, and Nan Yang. 2021. "Matrix stiffness modulates hepatic stellate cell activation into tumor-promoting myofibroblasts via E2F3-dependent signaling and regulates malignant progression." *Cell Death & Disease* 12 (12): 1134.

Mannaerts, I., S. B. Leite, S. Verhulst, S. Claerhout, N. Eysackers, L. F. Thoen, A. Hoorens, H. Reynaert, G. Halder, and L. A. van Grunsven. 2015. "The Hippo pathway effector YAP controls mouse hepatic stellate cell activation." *J Hepatol* 63 (3): 679-88.

Mormone, Elisabetta, Joseph George, and Natalia Nieto. 2011. "Molecular pathogenesis of hepatic fibrosis and current therapeutic approaches." *Chemico-Biological Interactions* 193 (3): 225-231.

Olsen, A. L., S. A. Bloomer, E. P. Chan, M. D. Gaça, P. C. Georges, B. Sackey, M. Uemura, P. A. Janmey, and R. G. Wells. 2011. "Hepatic stellate cells require a stiff environment for myofibroblastic differentiation." *Am J Physiol Gastrointest Liver Physiol* 301 (1): G110-8.

Ortiz, Cristina, Robert Schierwagen, Liliana Schaefer, Sabine Klein, Xavier Trepat, and Jonel Trebicka. 2021. "Extracellular Matrix Remodeling in Chronic Liver Disease." *Current Tissue Microenvironment Reports* 2 (3): 41-52.

Pei, R., T. Si, Y. Lu, J. X. Zhou, and L. Jiang. 2018. "Salvianolic acid A, a novel PI3K/Akt inhibitor, induces cell apoptosis and suppresses tumor growth in acute myeloid leukemia." *Leuk Lymphoma* 59 (8): 1959-1967.

Pellicoro, Antonella, Prakash Ramachandran, John P. Iredale, and Jonathan A. Fallowfield. 2014. "Liver fibrosis and repair: immune regulation of wound healing in a solid organ." *Nature Reviews Immunology* 14 (3): 181-194.

Penke, L. R., S. K. Huang, E. S. White, and M. Peters-Golden. 2014. "Prostaglandin E2 inhibits α-smooth muscle actin transcription during myofibroblast differentiation via distinct mechanisms of modulation of serum response factor and myocardin-related transcription factor-A." *J Biol Chem* 289 (24): 17151-62.

Rajakylä, E. K., and M. K. Vartiainen. 2014. "Rho, nuclear actin, and actin-binding proteins in the regulation of transcription and gene expression." *Small GTPases* 5: e27539.

Rajgor, D., and C. M. Shanahan. 2013. "Nesprins: from the nuclear envelope and beyond." *Expert Rev Mol Med* 15: e5.

Renga, Barbara, Andrea Mencarelli, Marco Migliorati, Sabrina Cipriani, Claudio D'Amore, Eleonora Distrutti, and Stefano Fiorucci. 2011. "SHP-dependent and -independent induction of peroxisome proliferator-activated receptor-γ by the bile acid sensor farnesoid X receptor counter-regulates the pro-inflammatory phenotype of liver myofibroblasts." *Inflammation Research* 60 (6): 577-587.

Rockey, D. C., Q. Du, N. D. Weymouth, and Z. Shi. 2019. "Smooth Muscle α-Actin Deficiency Leads to Decreased Liver Fibrosis via Impaired Cytoskeletal Signaling in Hepatic Stellate Cells." *Am J Pathol* 189 (11): 2209-2220.

Rockey, Don C., Nate Weymouth, and Zengdun Shi. 2013. "Smooth Muscle α Actin (Acta2) and Myofibroblast Function during Hepatic Wound Healing." *PLOS ONE* 8 (10): e77166.

Rombouts, K., T. Mello, F. Liotta, A. Galli, A. Caligiuri, F. Annunziato, and M. Pinzani. 2012. "MARCKS actin-binding capacity mediates actin filament assembly during mitosis in human hepatic stellate cells." *Am J Physiol Cell Physiol* 303 (4): C357-67.

Rumgay, Harriet, Melina Arnold, Jacques Ferlay, Olufunmilayo Lesi, Citadel J. Cabasag, Jérôme Vignat, Mathieu Laversanne, Katherine A. McGlynn, and Isabelle Soerjomataram. 2022. "Global burden of primary liver cancer in 2020 and predictions to 2040." *Journal of Hepatology* 77 (6): 1598-1606.

Seo, J., and J. Kim. 2018. "Regulation of Hippo signaling by actin remodeling." *BMB Rep* 51 (3): 151-156.

Shi, Z., M. Ren, and D. C. Rockey. 2020. "Myocardin and myocardin-related transcription factor-A synergistically mediate actin cytoskeletal-dependent inhibition of liver fibrogenesis." *Am J Physiol Gastrointest Liver Physiol* 318 (3): G504-g517.

Shimada, Hideaki, Nicholas R. Staten, and Lakshman E. Rajagopalan. 2011. "TGF-β1 mediated activation of Rho kinase induces TGF-β2 and endothelin-1 expression in human hepatic stellate cells." *Journal of Hepatology* 54 (3): 521-528.

Sicklick, Jason K., Yin-Xiong Li, Steve S. Choi, Yi Qi, Wei Chen, Marcia Bustamante, Jiawen Huang, Marzena Zdanowicz, Terese Camp, Michael S. Torbenson, Marcos Rojkind, and Anna Mae Diehl. 2005. "Role for Hedgehog signaling in hepatic stellate cell activation and viability." *Laboratory Investigation* 85 (11): 1368-1380.

Stalnikowitz, David Kershenobich, and Alan Bonder Weissbrod. 2003. "Liver fibrosis and inflammation. A review." *Annals of Hepatology* 2 (4): 159-163.

Stefano, Fiorucci, Rizzo Giovanni, Antonelli Elisabetta, Renga Barbara, Mencarelli Andrea, Riccardi Luisa, Morelli Antonio, Pruzanski Mark, and Pellicciari Roberto. 2005. "Cross-Talk between Farnesoid-X-Receptor (FXR) and Peroxisome Proliferator-Activated Receptor γ Contributes to the Antifibrotic Activity of FXR Ligands in Rodent Models of Liver Cirrhosis." *Journal of Pharmacology and Experimental Therapeutics* 315 (1): 58.

Tacke, Frank, and Ralf Weiskirchen. 2012. "Update on hepatic stellate cells: pathogenic role in liver fibrosis and novel isolation techniques." *Expert Review of Gastroenterology & Hepatology* 6 (1): 67-80.

Tsuchida, Takuma, and Scott L. Friedman. 2017. "Mechanisms of hepatic stellate cell activation." *Nature Reviews Gastroenterology & Hepatology* 14 (7): 397-411.

Vardouli, Lina, Eleftheria Vasilaki, Elsa Papadimitriou, Dimitris Kardassis, and Christos Stournaras. 2008. "A novel mechanism of TGFβ-induced actin reorganization mediated by Smad proteins and Rho GTPases." *The FEBS Journal* 275 (16): 4074-4087.

Villalonga, P., and A. J. Ridley. 2013. "Rho GTPases and Actin Cytoskeleton Dynamics." In *Encyclopedia of Biological Chemistry (Second Edition)*, edited by William J. Lennarz and M. Daniel Lane, 111-115. Waltham: Academic Press.

Wang, Jiaxu, Ron Zohar, and Christopher A. McCulloch. 2006. "Multiple roles of α-smooth muscle actin in mechanotransduction." *Experimental Cell Research* 312 (3): 205-214.

Wang, R., F. Song, S. Li, B. Wu, Y. Gu, and Y. Yuan. 2019. "Salvianolic acid A attenuates CCl(4)-induced liver fibrosis by regulating the PI3K/AKT/mTOR, Bcl-2/Bax and caspase-3/cleaved caspase-3 signaling pathways." *Drug Des Devel Ther* 13: 1889-1900.

Wells, R. G., and D. E. Discher. 2008. "Matrix elasticity, cytoskeletal tension, and TGF-beta: the insoluble and soluble meet." *Sci Signal* 1 (10): pe13.

Wen, Dongsheng, Ya Gao, Chiakang Ho, Li Yu, Yuguang Zhang, Guozhong Lyu, Dahai Hu, Qingfeng Li, and Yifan Zhang. 2022. "Focusing on Mechanoregulation Axis in Fibrosis: Sensing, Transduction and Effecting." *Frontiers in Molecular Biosciences* 9.

Wickstead, B., and K. Gull. 2011. "The evolution of the cytoskeleton." *J Cell Biol* 194 (4): 513-25.

Wipff, Pierre-Jean, and Boris Hinz. 2008. "Integrins and the activation of latent transforming growth factor β1 – An intimate relationship." *European Journal of Cell Biology* 87 (8): 601-615.

Xu, Quanchen, Yuanxin Miao, Jizheng Ren, Yu Sun, Cong Li, Xia Cai, and Zhiguo Wang. 2022. "Silencing of Nesprin-2 inhibits the differentiation of myofibroblasts from fibroblasts induced by mechanical stretch." *International Wound Journal* 19 (5): 978-986.

Yan, Yufei, Jiefei Zeng, Linhao Xing, and Changyong Li. 2021. "Extra- and Intra-Cellular Mechanisms of Hepatic Stellate Cell Activation." *Biomedicines* 9: 1014.

Yee, H. F., Jr. 1998. "Rho directs activation-associated changes in rat hepatic stellate cell morphology via regulation of the actin cytoskeleton." *Hepatology* 28 (3): 843-50.

Yu, Tingting, Ke Hu, Qiufan Xu, Xueqin Xu, Chunyue Du, Yinyi Zhao, Sen Yan, Wei Wang, Shijia Tang, Shen Yue, Xiya Zhang, Danlei Wang, Feimin Zhang, Steven Yan Cheng, and Ning Gu. 2020. "An Easy-to-Fabricate Hydrogel Platform with Tunable Stiffness and Cell Anchorage: Validation of Its Feasibility in Modulating Sonic Hedgehog Signaling Pathway Physically." *Macromolecular Materials and Engineering* 305 (4): 1900759.

Zhang, J., N. Jiang, J. Ping, and L. Xu. 2021. "TGF-β1-induced autophagy activates hepatic stellate cells via the ERK and JNK signaling pathways." *Int J Mol Med* 47 (1): 256-266.

Chapter 8

Audiovestibular Disorders in Children with Congenital Cytomegalovirus Infections

Virginia Corazzi
Cecilia Albi
Claudio Moretti
Chiara Bianchini
and Andrea Ciorba*
ENT & Audiology Unit, Department of Neurosciences and Rehabilitation, University Hospital of Ferrara, Italy

Abstract

Cytomegalovirus (CMV) is nowadays the most frequent cause of congenital infection worldwide. About 0.6 to 6.1 per cent of children are born every year with congenital CMV infection (cCMV) [1], representing a major social and health burden. Actually, cCMV infection is the leading cause of permanent non-genetic sensorineural hearing loss (SNHL) in children and the most frequent infectious cause of neurological disorders [2, 3]. CMV has an elective tropism for the cochlea and posterior labyrinth as demonstrated by histopathological studies. cCMV specifically accounts for 21% of SNHL at birth in developed nations [4], while the vestibular impairment can affect a posturomotor development delay, particularly if bilateral and complete. The symptomatic form of cCMV can be diagnosed by clinical manifestations [5-7]; SNHL is present in about half of the symptomatic cases and may be detected through the neonatal hearing screening

* Corresponding Author's Email: andrea.ciorba@unife.it.

In: Advances in Biology. Volume 4
Editor: Charles D. Grant
ISBN: 979-8-88697-999-2

program. Conversely, the asymptomatic children still remain underestimated, due to the absence of a universal neonatal CMV screening. To date, there is no effective curative treatment available for CMV-related hearing impairment. The treatment with ganciclovir or valganciclovir in symptomatic forms of cCMV has been demonstrated to determine a stabilization or even an improvement of the hearing threshold [8-11]. Nonetheless, in case of early identification of hearing loss, a prompt auditory habilitation/rehabilitation with a cochlear implant or hearing aid, when indicated, represents the conventional strategy to promote adequate language development in children suffering from cCMV. Moreover, antiviral therapy is currently not recommended for the treatment of children presenting with asymptomatic cCMV and isolated SNHL. Eventually, an approved CMV vaccine is currently lacking, even if numerous promising clinical trials are still ongoing.

Aim of this chapter is to describe the audiovestibular disorders in children with cCMV, also particularly focusing on the early detection strategies and current rehabilitation-therapeutic options, according to the most recent literature findings.

Keywords: congenital cytomegalovirus infection, sensorineural hearing loss, vestibular impairment, ganciclovir, valganciclovir

Introduction

Cytomegalovirus (CMV) is a virus belonging to the herpesvirus family, which is globally endemic and has humans as its only host. The virus is transmitted by individuals with primary infection or reactivation of the infection through body fluids such as blood, saliva, urine, tears, seminal fluid, vaginal secretion, and milk [12]. While in immunocompetent individuals acquired CMV infection manifests itself with mild flu-like symptoms, congenital CMV (cCMV) infection can manifest itself with severe audiological, neurological, and cognitive sequelae [2, 3]. Maternal-infant transmission can occur in the pre-natal period by transplacental route, but it can also occur during the peri-natal period through vaginal secretions or again in the post-natal period through breast milk. The risk of permanent sensorineural damages is greatest when the infection is primarily acquired by the mother during the first trimester of pregnancy [13].

Clinical Presentation of Congenital Cytomegalovirus Infection

cCMV infection may present as a symptomatic or asymptomatic disease. Among affected infants only 10% present at birth with symptoms, while among asymptomatic newborns approximately 15% may develop late sequelae of the infection [4], namely permanent neurological disorders, psychomotor developmental delays, cognitive defects, and hearing loss [2, 3]. At birth, infants with symptomatic cCMV infection may present with jaundice, thrombocytopenia and neutropenia, hepatosplenomegaly, petechiae, and purpura. Furthermore, these infants are often low birth weighing and have head circumference below the 10th percentile [14]. The presence of microcephaly is considered a prognostic factor strongly correlated with the development of long-term sensorineural sequelae, particularly when associated with intrauterine growth restriction [15, 16].

CMV has an elective tropism for the cochlea and posterior labyrinth as demonstrated by histopathological studies [17]. Actually, cCMV infection is the leading cause of permanent non-genetic sensorineural hearing loss (SNHL) in children [2, 3], and specifically accounts for 21% of SNHL at birth in developed nations [4]. The most frequent audiological manifestation of cCMV infection is unilateral or bilateral mild-to-profound pantonal SNHL and its clinical presentation can be extremely heterogeneous [18, 19]. In fact, SNHL can be present at birth, or show itself with a late onset, also with a progressive or fluctuant trend [18, 19].

Vestibular impairment induced by cCMV infection may be associated with hearing loss or occasionally be isolated. However, the real prevalence of vestibular involvement in children with cCMV infection is probably still underestimated due to the difficulties of performing postural and vestibular assessment in little patients and possibly also because the vestibular symptoms may easily overlap with neurological sequelae, often co-present in the disease CMV-related [20, 21]. Vestibular impairment could induce hypotonia and poor balance [22]. Thus, the involvement of the posterior labyrinth can lead to posturomotor development delay, especially when bilateral and complete. In fact, bilateral vestibular impairment can cause deficits in dynamic vision and impaired perception of space and gravity leading to dyspraxia and motor learning difficulties. Also, dizziness and/or vertigo have been reported in infants and children suffering from cCMV infection [21]; in particular, episodic vertigo with neurovegetative symptoms has been attributed to delayed development of endolymphatic hydrops after cCMV infection [23-25].

Audiological Assessment in Congenital Cytomegalovirus Infection

"Targeted" newborn screening for CMV infection is a protocol aiming at the early identification of cCMV infection in selected cases [26]. In particular, nowadays, it is offered to all infants who have failed the Newborn Hearing Screening (NHS) at birth. Thus, the detection of CMV DNA in newborn urine and saliva specimens should be carried out within the first 2-3 weeks of life by real-time Polymerase Chain Reaction (PCR) technique [27] whenever even unilateral OAE (otoacoustic emission) and/or AABR (automated auditory brainstem response) "fail" result occurs at NHS. The audiological assessment should continue referring children affected by cCMV infection and suspected hearing loss to an audiologist for rescreening and, when indicated, for a comprehensive evaluation within three months of life. The hearing threshold could be initially estimated by clinical ABR and Auditory Steady State Response (ASSR), while behavioural audiometry could be performed as the child grows [28]. Due to the possible long-term development of SNHL, audiological follow-up should be extended until 4-6 years of age in children with cCMV infection, especially because the reported threshold elevation is attested between 3 and 4 years of age [3, 29].

Since the majority of newborns with cCMV infection have normal hearing at birth but remain at risk of developing late onset SNHL, a long-term audiological follow-up is strongly recommended even in infected children with normal hearing for prompt identification of hearing loss development [4]. Unfortunately, the asymptomatic children still remain underestimated, due to the absence of a universal neonatal CMV screening; thus, in these children, hearing defects could possibly remain undetected and untreated, with negative consequences for these little patients [4].

Vestibular and Postural Assessment in Congenital Cytomegalovirus Infection

Vestibular and postural assessment is important especially in presence of posture-motor defects, recurring falls, and delayed walking, in order to differentiate in children with cCMV infection among vestibular impairment and neurological diseases / oculomotor disorders [20]. Nonetheless, the differential diagnosis could be very difficult in these children, due to the frequent co-presence of cerebellar disorders, cerebral palsy, central hypotonia, convulsions, autism spectrum disorder, and attention deficit disorder. In fact, oculomotricity assessment and neurological/neurodevelopmental testing are recommended together with vestibular and postural assessment for the differential diagnosis, when possible [21].

Some standing balance tasks can be performed in order to assess postural-control impairment in collaborating children, such as the Romberg manoeuvre, the tandem Romberg, and the march in place [30]. Also standardized tests for motor development and balance validated for the paediatric population can be used, such as the Movement ABC-2, the Bruininks-Oseretsky Test of Motor Proficiency (second edition, BOT-2), the Peabody Developmental Motor Scales – second edition (PDMS-2), the Alberta Infant Motor Scales (AIMS), and the Ghent Developmental Balance Test (GDBT) [20, 31]. In even smaller patients, simple behavioural observation during walking or playing could provide important evidence about the motor-milestone acquisition. Furthermore, information obtained from the parental interview could offer further data about the postural skills of the very little patients.

Vestibular assessment in cCMV children includes tests for otolith function examination, such as cervical Vestibular Evoked Myogenic Potentials (cVEMPs), ocular VEMPs (oVEMPs), and OVAR (off-vertical axis rotation), and for canal function examination, such as the caloric exam, the video Head Impulse test (vHIT), and the earth-vertical axis rotation test (EVAR), depending on the age and collaboration of the little patients [20, 21].

Prevention Strategies and Medical Treatment for Congenital Cytomegalovirus Infection

Primary Prevention

Despite the efforts over the recent decades, the development of an effective vaccine has not yet occurred [32]. The possible main reasons at the basis of the difficulty to develop an anti-CMV vaccine could be: i) the significant genetic diversity of the CMV family, ii) CMV's ability to evade the immune defences that make CMV re-infections very common, and iii) the paucity of experimental animal models [33]. Although several promising clinical trials are still ongoing, among the most promising could be counted the mRNA vaccines, the Adjuvanted Recombinant Protein Vaccine, and the DNA vaccines. mRNA-1647 (which induces responses against glycoprotein B and the pentamer gHgLpUL128L) is currently in a phase III study, having demonstrated high immunogenic potential and tolerability in previous studies [34]. Adjuvanted Recombinant Protein Vaccines have been studied in phase I and II trials showing nearly 50% seroconversion ability in women of childbearing age with negative serology for CMV [35, 36]. Currently in phase III trial, DNA vaccines are composed of plasmids encoding p65 and glycoprotein B, characterized by the ability to stimulate both T-mediated response and antibody production [37]. Pending the definitive development of a vaccine, the only primary prevention rule that can be safely applied to date is reported to be the proper counselling of women of childbearing age and pregnant women and hygiene standards (such as washing hands regularly) to minimize the possibility of infection [38].

Secondary Prevention

The goal of secondary prevention is to reduce maternal-foetal transmission of the disease once the mother's seroconversion is evidenced. Studies have mainly focused on two different methods: hyperimmune globulins (HIG) and antiviral drugs (AVD) [39, 40]. As for HIG, current scientific evidence is discordant: in a multicentre, double-blind trial published in the New England Journal of Medicine in 2021 conducted on 399 CMV-seropositive pregnant women, it was shown that the administration of HIG was not effective in reducing maternal-foetal transmission, symptom onset, and mortality [41]. Nevertheless, less recent studies had shown that administration of HIG in CMV-positive pregnant women reduced maternal-foetal transmission [42] and improved neurological and audiological outcomes [43]. More robust, however, is the evidence regarding the administration of AVD; specifically, the administration of valaciclovir, especially when initiated in the periconceptional period or in the first trimester, showed a 70% reduction in the risk of transmission [44]. Other AVD such as Ganciclovir and Letamir are still being studied [45].

Tertiary Prevention

Tertiary prevention aims to reduce symptomatic presentation after establishing foetal infection. Prenatal use of ultrasound (US), magnetic resonance imaging (MRI), and foetal blood sampling can be used to assess the prognosis of the foetus [46-48]. If imaging changes suggest a good to moderate prognosis, transplacental therapy with valacyclovir is now being investigated, as demonstrated in a clinical trial that showed good tolerability and improved long-term outcomes [49].

Medical Treatment in Children with Congenital Cytomegalovirus Infection

According to the most recent recommendations [5, 6], antiviral drugs are offered to infants born after 32 weeks of gestation only in case of moderate-to-severe symptomatic cCMV infection.

Antiviral drugs administration has been shown to be associated with a reduction in developing complications, particularly those of audiological concern [50]. Intravenous ganciclovir (6 mg/kg, 2 times per day) or oral valganciclovir (16 mg/kg, 2 times per day) should be administrated within the first month of life [51, 52]. In case of congenital hearing loss, it has been reported that these antiviral drugs can improve or stabilize the hearing threshold [50]. Even so, due to the potential toxicities of these drugs, such as myelosuppression (associated with neutropenia, anaemia, and thrombocytopenia) and carcinogenicity [5, 6, 51, 52], this medical treatment is not currently recommended in children with neither asymptomatic nor symptomatic cCMV infection and isolated hearing loss [5, 6].

Audiovestibular Rehabilitation Strategies

Audiovestibular rehabilitation interventions in children with cCMV infection are crucial to foster the growth of communication, language, and balance skills, which facilitate social integration.

Audiological Rehabilitation

SNHL remains the most common complication in both symptomatic (one in two) and asymptomatic (one in ten) cCMV infection [53], so care and audiological follow-up are essential for early diagnosis and definition of the rehabilitation strategies [4]. While for moderate degrees of SNHL, the use of hearing aids remains to date the conventional intervention, for severe and profound SNHL, a cochlear implant (CI) represents the most effective means of audiological rehabilitation/habilitation to promote adequate language development [54], even in those cases with neurological and/or cognitive deficits [55, 56]. In this scenario, the speech therapist has a fundamental role in developing language skills, assessing perceptual abilities, and vocabulary expansion over the follow-up [57].

Vestibular and Postural Rehabilitation

Vestibular sequelae in children with cCMV infection are frequent but often misdiagnosed, especially in presence of neurological comorbidities. Early diagnosis of vestibular impairment is of paramount importance because it allows an early rehabilitation pathway to be established to reduce the burden on the child's growth [58]. When vestibular impairment is identified, proper vestibular and postural rehabilitation should begin before the age of 2 years and be carried out with the collaboration of a physiotherapist. Exercises that increase somatosensory and visual awareness should be set up with fine motor exercises, such as hand-eye coordination exercises increasing visual stability and general balance [21, 31].

Conclusion

To date, cCMV infection represents one of the major health problems worldwide given its high prevalence and its association with major complications, particularly SNHL. Early diagnosis of cCMV is associated with a greater likelihood of reducing the occurrence of long-term sequelae of infection [4, 59]. While waiting for a safe, effective, and tolerated vaccine, counselling to fertile and pregnant women, adherence to hygiene standards, and increased public attention to this problem represent the most effective form of primary prevention [60]. To maximise functional results in affected children, early identification, multidisciplinary teamwork, and evidence-based therapeutic-rehabilitation strategies are crucial. Currently, a complete audiovestibular assessment follow-up and management should be recommended in all children affected by cCMV infection or, at least, in all cCMV children presenting SNHL. For the near future, there are open issues to be addressed: i) to develop effective tools to detect women at increased risk of CMV infection; ii) to implement cost-effective strategies for the early diagnosis of asymptomatic cCMV infection; iii) to implement screening tests for detecting vestibular disorders in infected children; iv) to better define specific medical therapeutic and targeted protocols for infants with both symptomatic and asymptomatic cCMV infection and hearing loss, and especially in case of late onset SNHL; v) to estimate long term effects on hearing function of antiviral drugs.

Disclaimer

None

References

[1] Lanzieri, T. M., Dollard, S. C., Bialek, S. R. and Grosse, S. D. (2014). Systematic review of the birth prevalence of congenital cytomegalovirus infection in developing countries. *International Journal of Infectious Diseases*, 22: 44-48. doi:10.1016/j.ijid.2013.12.010.

[2] Morton, C. C. and Nance, W.E. (2006). Newborn hearing screening--a silent revolution. *The New England Journal of Medicine*, 354 (20): 2151-2164. doi:10.1056/NEJMra050700.

[3] Dahle, A. J., Fowler, K. B., Wright, J. D., Boppana, S. B., Britt, W. J. and Pass, R. F. (2000). Longitudinal investigation of hearing disorders in children with congenital cytomegalovirus. *Journal of the American Academy of Audiology*, 11 (5): 283-290.

[4] Manicklal, S., Emery, V. C., Lazzarotto, T., Boppana, S. B. and Gupta, R. K. (2013). The "silent" global burden of congenital cytomegalovirus. *Clinical Microbiology Reviews*, 26 (1): 86-102

[5] Luck, S. E., Wieringa, J. W., Blázquez-Gamero, D., Henneke, P., Schuster, K., Butler, K., Capretti, M. G., Cilleruelo, M. J., Curtis, N., Garofoli, F., Heath, P., Iosifidis, E., Klein, N., Lombardi, G., Lyall, H., Nieminen, T., Pajkrt, D., Papaevangelou, V., Posfay-Barbe, K., Puhakka, L., Roilides, E., Rojo, P., Saavedra-

Lozano, J., Shah, T., Sharland, M., Saxen, H., Vossen, A. C. T. M. and ESPID Congenital CMV Group Meeting, Leipzig 2015 (2017). Congenital Cytomegalovirus: A European Expert Consensus Statement on Diagnosis and Management. *The Pediatric Infectious Disease Journal*, 36 (12): 1205-1213. doi:10.1097/INF.0000000000001763.

[6] Rawlinson, W. D., Boppana, S. B., Fowler, K. B., Kimberlin, D. W., Lazzarotto, T., Alain, S., Daly, K., Doutré, S., Gibson, L., Giles, M. L., Greenlee, J., Hamilton, S. T., Harrison, G. J., Hui, L., Jones, C. A., Palasanthiran, P., Schleiss, M. R., Shand, A. W. and van Zuylen W. J. (2017). Congenital cytomegalovirus infection in pregnancy and the neonate: consensus recommendations for prevention, diagnosis, and therapy. *The Lancet Infectious Diseases*, 17 (6): e177-e188. doi:10.1016/S1473-3099(17)30143-3.

[7] Dietrich, M. L. and Schieffelin, J. S. (2019) Congenital Cytomegalovirus Infection. *Ochsner Journal*, 19 (2): 123-130.

[8] Harrison, G.J. (2015) Current controversies in diagnosis, management, and prevention of congenital cytomegalovirus: updates for the pediatric practitioner. *Pediatr. Annals*, 44 (5): e115-125.

[9] James, S.H. and Kimberlin, D.W. (2016) Advances in the prevention and treatment of congenital cytomegalovirus infection. *Current Opinion in Pediatrics*, 28 (1): 81-85

[10] Kimberlin, D. W., Lin, C. Y., Sánchez, P. J., Demmler, G. J., Dankner, W., Shelton, M., Jacobs, R. F., Vaudry, W., Pass, R. F., Kiell, J. M., Soong, S. J., Whitley, R. J. and National Institute of Allergy and Infectious Diseases Collaborative Antiviral Study Group (2003). Effect of ganciclovir therapy on hearing in symptomatic congenital cytomegalovirus disease involving the central nervous system: a randomized, controlled trial. *The Journal of Pediatrics*, 143: 16-25

[11] Kimberlin, D. W., Jester, P. M., Sánchez, P. J., Ahmed, A., Arav-Boger, R., Michaels, M. G., Ashouri, N., Englund, J.A., Estrada, B., Jacobs, R. F., Romero, J. R., Sood, S. K., Whitworth, M. S., Abzug, M. J., Caserta, M. T., Fowler, S ., Lujan-Zilbermann, J., Storch, G. A., DeBiasi, R. L., Han, J. Y., Palmer, A., Weiner, L. B., Bocchini, J. A., Dennehy, P. H., Finn, A., Griffiths, P. D., Luck, S., Gutierrez, K., Halasa, N., Homans, J., Shane, A. L., Sharland, M., Simonsen, K., Vanchiere, J. A., Woods, C. R., Sabo, D. L., Aban, I., Kuo, H., James, S. H., Prichard, M. N., Griffin, J., Giles, D., Acosta, E. P., Whitley, R. J. and National Institute of Allergy and Infectious Diseases Collaborative Antiviral Study Group. (2015). Valganciclovir for symptomatic congenital cytomegalovirus disease. *The New England Journal of Medicine*, 372 (10): 933-943.

[12] Leruez-Ville, M., Foulon, I., Pass, R. and Ville, Y. (2020). Cytomegalovirus infection during pregnancy: state of the science. *American Journal of Obstetrics & Gynecology*, 223 (3): 330–349. doi: 10.1016/j.ajog.2020.02.018.

[13] Pass, R. F., Fowler, K. B., Boppana, S. B., Britt, W. J. and Stagno, S. (2006). Congenital cytomegalovirus infection following first trimester maternal infection: symptoms at birth and outcome. *Journal of Clinical Virology*, 35: 216-220. doi: 10.1016/j.jcv.2005.09.015.

[14] Boppana, S. B., Ross, S. A. and Fowler, K. B. (2013). Congenital cytomegalovirus infection: clinical outcome. *Clinical Infectious Diseases*, 57: S178–S181. doi:10.1093/cid/cit629.

[15] Stagno, S., Dworsky, M. E., Torres, J., Mesa, T. and Hirsh, T. (1982). Prevalence and importance of congenital cytomegalovirus infection in three different populations. *The Journal of Pediatrics*, 101 (6): 897–900. doi:10.1016/s0022-3476(82)80006-1

[16] Rivera, L. B., Boppana, S. B., Fowler, K. B., Britt, W. J., Stagno, S. and Pass, R. F. (2002). Predictors of hearing loss in children with symptomatic congenital cytomegalovirus infection. *Pediatrics*, 110 (4): 762–767. doi:10.1542/peds.110.4.762

[17] Teissier, N., Delezoide, A. L., Mas, A. E., Khung-Savatovsky, S., Bessières, B., Nardelli, J., Vauloup-Fellous, C., Picone, O., Houhou, N., Oury, J.F., Van Den Abbeele, T., Gressens, P. and Adle-Biassette, H. (2011). Inner ear lesions in congenital cytomegalovirus infection of human fetuses. *Acta Neuropathologica*, 122 (6): 763 -774. doi: https://doi.org/10.1007/s00401-011-0895-y.

[18] Ross, S. A., Fowler, K. B., Ashrith, G., Stagno, S., Britt, W. J., Pass, R. F. and Boppana, S. B. (2006). Hearing loss in children with congenital cytomegalovirus infection born to mothers with preexisting immunity. *The Journal of Pediatrics*, 148 (3): 332–336. doi:10.1016/j.jpeds.2005.09.003.

[19] Cohen, B. E., Durstenfeld, A. and Roehm, P. C. (2014). Viral causes of hearing loss: a review for hearing health professionals. *Trends in Hearing*, 18, 2331216514541361. doi:10.1177/2331216514541361.

[20] Corazzi, V., Hatzopoulos, S., Bianchini, C., Skarżyńska, M.B., Pelucchi, S., Skarżyński, P.H. and Ciorba, A. (2022). Vestibular and postural impairment in congenital Cytomegalovirus infection. *International Journal of Pediatric Otorhinolaryngology*, 152: 111005. doi: 10.1016/j.ijporl.2021.111005.

[21] Bernard, S., Wiener-Vacher, S., Van Den Abbeele, T. and Teissier, N. (2015). Vestibular Disorders in Children With Congenital Cytomegalovirus Infection. *Pediatrics*, 136 (4): e887–e895. doi:10.1542/peds.2015-0908.

[22] Shears, A., Yan, G., Mortimer, H., Cross, E., Sapuan, S., Kadambari, S., Luck, S., Heath, P. T., Walter, S. and Fidler, K. J. (2022). Vestibular and balance dysfunction in children with congenital CMV: a systematic review. *Archives of Disease in Childhood. Fetal and Neonatal Edition*, 107 (6): 630–636. doi:10.1136/archdischild-2021-323380.

[23] Weiss J. S. and Ronis M. L. (1977). The otologic manifestations of cytomegalic inclusion disease - a case report. *Transactions - Pennsylvania Academy of Ophthalmology and Otolaryngology*, 30: 52–54.

[24] Huygen, P. L. and Admiraal R. J. (1996). Audiovestibular sequelae of congenital cytomegalovirus infection in 3 children presumably representing 3 symptomatically different types of delayed endolymphatic hydrops. *International Journal of Pediatric Otorhinolaryngology*, 35: 143–154. doi: 10.1016/0165-5876(96)83899-7.

[25] Dhondt, C., Maes, L., Oostra, A. and Dhooge I. (2019). Episodic vestibular symptoms in children with a congenital cytomegalovirus infection: a case series. *Otology & Neurotology*, 40: e636–e642. doi:10.1097/MAO.0000000000002244.

[26] Gantt, S., Dionne, F., Kozak, F. K., Goshen, O., Goldfarb, D. M., Park, A. H., Boppana, S.B. and Fowler K. (2016). Cost-effectiveness of universal and targeted newborn screening for congenital cytomegalovirus infection. *JAMA Pediatrics*, 170 (12): 1173-1180. doi: 10.1001/jamapediatrics.2016.2016.

[27] Fowler, K. B., McCollister, F. P., Sabo, D. L., Shoup, A. G., Owen, K. E., Woodruff, J. L., Cox, E., Mohamed, L.S., Choo, D.I. and Boppana S.B. (2017). CHIMES Study. A Targeted Approach for Congenital Cytomegalovirus Screening Within Newborn Hearing Screening. *Pediatrics*, 139 (2): e20162128. doi: 10.1542/peds.2016-2128.

[28] Harlor, A. D. and Bower, C. (2009). Committee on Practice and Ambulatory Medicine; Section on Otolaryngology-Head and Neck Surgery. Hearing assessment in infants and children: recommendations beyond neonatal screening. *Pediatrics*, 124 (4): 1252-1263. doi: 10.1542/peds.2009-1997.

[29] Fowler, K. B., McCollister, F. P., Dahle, A. J., Boppana, S., Britt, W. J. and Pass, R. F. (1997). Progressive and fluctuating sensorineural hearing loss in children with asymptomatic congenital cytomegalovirus infection. *The Journal of Pediatrics*, 130 (4): 624–630. doi:10.1016/s0022-3476(97)70248-8.

[30] Valente, L. M. (2011). Assessment techniques for vestibular evaluation in pediatric patients. *Clinical Otolaryngology*, 44 (2): 273–290. doi: 10.1016/j.otc.2011.01.002.

[31] Rine, R. M. and Wiener-Vacher, S. (2013). Evaluation and treatment of vestibular dysfunction in children. *NeuroRehabilitation*, 32 (3): 507–518. doi:10.3233/NRE-130873.

[32] Scarpini, S., Morigi. F., Betti, L., Dondi, A., Biagi, C. and Lanari, M. (2021). Development of a Vaccine against Human Cytomegalovirus: Advances, Barriers, and Implications for the Clinical Practice. *Vaccines*, 9 (6): 551. doi: 10.3390/vaccines9060551.

[33] Inoue, N., Abe, M., Kobayashi, R. and Yamada S. (2018). Vaccine Development for Cytomegalovirus. Advances in Experimental Medicine and Biology, 1045: 271-296. doi: 10.1007/978-981-10-7230-7_13.

[34] Manus, J. M. (2022). Un vaccin à ARNm contre le cytomégalovirus [An mRNA vaccine against cytomegalovirus]. *Revue Francophone des Laboratoires*, (538): 10-11. doi: 10.1016/S1773-035X(21)00350-6.

[35] Bernstein, D.I., Munoz, F.M., Callahan, S.T., Rupp, R., Wootton, S.H., Edwards, K.M., Turley, C.B., Stanberry, L.R., Patel, S.M. and McNeal, M.M. (2016). Safety and efficacy of a cytomegalovirus glycoprotein B (gB) vaccine in adolescent girls: A randomized clinical trial. *Vaccine*, 34 (3): 313–319. doi: 10.1016/ j.vaccine.2015.11.056.

[36] Pass, R. F., Zhang, C., Evans, A., Simpson, T., Andrews, W., Huang, M. L., Corey, L., Hill, J., Davis, E. and Flanigan, C. (2009). Vaccine prevention of maternal cytomegalovirus infection. *The New England Journal of Medicine*, 360 (12): 1191–1199. doi: 10.1056/NEJMoa0804749.

[37] Hu, X., Wang, H. Y., Otero, C.E., Jenks, J. A. and Permar, S. R. (2022). Lessons from Acquired Natural Immunity and Clinical Trials to Inform Next-Generation Human Cytomegalovirus Vaccine Development. *Annual Review of Virology*, 9 (1): 491–520. doi: 10.1146/annurev-virology-100220-010653.

[38] Adler, S. P., Finney, J. W., Manganello, A. M. and Best, A. M. (1996). Prevention of child-to-mother transmission of cytomegalovirus by changing behaviors: a randomized controlled trial. *The Pediatric Infectious Disease Journal*, 15 (3): 240-246. doi: 10.1097/00006454-199603000-00013.

[39] Kagan, K. O., Enders, M., Hoopmann, M., Geipel, A., Simonini, C., Berg, C., Gottschalk, I., Faschingbauer, F., Schneider, M. O. and Ganzenmueller, T. (2021). Outcome of pregnancies with recent primary cytomegalovirus infection in first trimester treated with hyperimmunoglobulin: Observational study. *Ultrasound in Obstetrics & Gynecology*, 57 (4): 560–567. doi: 10.1002/uog.23596.

[40] Shahar-Nissan, K., Pardo, J., Peled, O., Krause, I., Bilavsky, E., Wiznitzer, A., Hadar, E. and Amir, J. (2020). Valaciclovir to prevent vertical transmission of cytomegalovirus after maternal primary infection during pregnancy: A randomised, double-blind, placebo-controlled trial. *Lancet*, 396 (10253): 779–785. doi: 10.1016/S0140-6736(20)31868-7.

[41] Hughes, B. L., Clifton, R. G., Rouse, D. J., Saade, G. R., Dinsmoor, M. J., Reddy, U. M., Pass, R., Allard, D., Mallett, G., Fette, L.M., Gyamfi-Bannerman, C., Varner, M. W., Goodnight, W. H., Tita, A. T. N., Costantine, M. M., Swamy, G. K., Gibbs, R. S., Chien, E. K., Chauhan, S. P., El-Sayed, Y. Y., Casey, B. M., Parry, S., Simhan, H.N., Napolitano, P.G., Macones, G.A. and Eunice Kennedy Shriver National Institute of Child Health and Human Development Maternal–Fetal Medicine Units Network. (2021). A Trial of Hyperimmune Globulin to Prevent Congenital Cytomegalovirus Infection. *The New England Journal of Medicine*, 385 (5): 436-444. doi: 10.1056/NEJMoa1913569.

[42] Kagan, K. O., Enders, M., Schampera, M. S., Baeumel, E., Hoopmann, M., Geipel, A., Berg, C., Goelz, R., De Catte, L., Wallwiener, D., Brucker, S., Adler, S.P., Jahn, G. and Hamprecht K. (2019). Prevention of maternal-fetal transmission of cytomegalovirus after primary maternal infection in the first trimester by biweekly hyperimmunoglobulin administration. *Ultrasound in Obstetrics & Gynecology*, 53 (3): 383-389. doi: 10.1002/uog.19164.

[43] Visentin, S., Manara, R., Milanese, L., Da Roit, A., Forner, G., Salviato, E., Citton, V., Magno, F.M., Orzan, E., Morando, C., Cusinato, R., Mengoli, C., Palu, G., Ermani, M., Rinaldi, R., Cosmi, E. and Gussetti N. (2012). Early primary cytomegalovirus infection in pregnancy: maternal hyperimmunoglobulin therapy improves outcomes among infants at 1 year of age. *Clinical Infectious Diseases*, 55 (4): 497-503. doi: 10.1093/cid/cis423.

[44] Shahar-Nissan, K., Pardo, J., Peled, O., Krause, I., Bilavsky, E., Wiznitzer, A., Hadar, E., Amir, J. (2020). Valaciclovir to prevent vertical transmission of cytomegalovirus after maternal primary infection during pregnancy: a randomised, double-blind, placebo-controlled trial. *Lancet*, 396 (10253): 779-785. doi: 10.1016/S0140-6736(20)31868-7.

[45] Faure Bardon, V., Peytavin, G., Le, M.P., Guilleminot, T., Elefant, E., Stirnemann, J., Leruez-Ville, M. and Ville, Y. (2020). Placental transfer of Letermovir & Maribavir in the ex vivo human cotyledon perfusion model. New perspectives for in utero treatment of congenital cytomegalovirus infection. *PLoS One*, 15 (4): e0232140. doi: 10.1371/journal.pone.0232140.

[46] Leruez-Ville, M., Stirnemann, J., Sellier, Y., Guilleminot, T., Dejean, A., Magny, J.F., Couderc, S., Jacquemard, F. and Ville, Y. (2016). Feasibility of predicting the outcome of fetal infection with cytomegalovirus at the time of prenatal diagnosis. *American Journal of Obstetrics & Gynecology*, 215 (3): 342.e1-9. doi: 10.1016/j.ajog.2016.03.052.

[47] Benoist, G., Salomon, L. J., Mohlo, M., Suarez, B., Jacquemard, F. and Ville, Y. (2008). Cytomegalovirus-related fetal brain lesions: Comparison between targeted ultrasound examination and magnetic resonance imaging. *Ultrasound in Obstetrics & Gynecology*, 32 (7): 900-905. doi: 10.1002/uog.6129.

[48] Lipitz, S., Yinon, Y., Malinger, G., Yagel, S., Levit, L., Hoffman, C., Rantzer, R. and Weisz, B. (2013). Risk of cytomegalovirus-associated sequelae in relation to time of infection and findings on prenatal imaging. *Ultrasound in Obstetrics & Gynecology*. 41 (5): 508-514. doi: 10.1002/uog.12377.

[49] Leruez-Ville, M., Ghout, I., Bussieres, L., Stirnemann, J., Magny, J.F., Couderc, S., Salomon, L.J., Guilleminot, T., Aegerter, P. and Benoist, G. (2016). In utero treatment of congenital cytomegalovirus infection with valacyclovir in a multicenter, open-label, phase II study. *American Journal of Obstetrics & Gynecology*, 215 (4): 462.e1-462.e10. doi: 10.1016/j.ajog.2016.04.003.

[50] Hamilton, S. T., van Zuylen, W., Shand, A., Scott, G.M., Naing, Z., Hall, B., Craig, M.E. and Rawlinson W.D. (2014). Prevention of congenital cytomegalovirus complications by maternal and neonatal treatments: a systematic review. *Reviews in Medical Virology*, 24 (6): 420-433. doi: 10.1002/rmv.1814.

[51] Mareri, A., Lasorella, S., Iapadre, G., Maresca, M., Tambucci, R. and Nigro, G. (2016). Anti-viral therapy for congenital cytomegalovirus infection: pharmacokinetics, efficacy and side effects. *Journal of Maternal-Fetal & Neonatal Medicine*, 29 (10): 1657-1664. doi: 10.3109/14767058.2015.1058774.

[52] Buonsenso, D., Serranti, D., Gargiullo, L., Ceccarelli, M., Ranno, O. and Valentini, P. (2012). Congenital cytomegalovirus infection: current strategies and future perspectives. *European Review for Medical and Pharmacological Sciences*, 16 (7): 919-935.

[53] Fowler, K. B. (2013). Congenital cytomegalovirus infection: audiologic outcome. *Clinical Infectious Diseases*, 57: S182-184. doi: 10.1093/cid/cit609.

[54] Bolduc, S. H., Bussières, R., Philippon, D. and Côté, M. (2021). The Correlation of Congenital CMV Infection and the Outcome of Cochlear Implantation. *The Journal of International Advanced Otology*, 17 (3): 190–194. doi: 10.5152/iao.2021.9335.

[55] Dobbie, A. M. (2017). Evaluation and management of cytomegalovirus associated congenital hearing loss. *Current Opinion in Otolaryngology & Head and Neck Surgery*, 25 (5): 390-395. doi: 10.1097/MOO.0000000000000401.

[56] Ciorba, A., Bovo, R., Trevisi, P., Bianchini, C., Arboretti, R. and Martini A. (2009). Rehabilitation and outcome of severe profound deafness in a group of 16 infants affected by congenital cytomegalovirus infection. *European Archives of Oto-Rhino-Laryngology*, 266 (10): 1539-1546. doi: 10.1007/s00405-009-0944-5.

[57] Viccaro, M., Filipo, R., Bosco, E., Nicastri, M., and Mancini, P. (2012). Long-term follow-up of implanted children with cytomegalovirus-related deafness. *Audiology and Neurotology*, 17 (6): 395–399. doi: 10.1159/000341160.

[58] Rine, R. M., Braswell, J., Fisher, D., Joyce, K., Kalar, K. and Shaffer, M. (2004). Improvement of motor development and postural control following intervention in children with sensorineural hearing loss and vestibular impairment. *International Journal of Pediatric Otorhinolaryngology*, 68 (9): 1141–1148. doi: 10.1016/j.ijporl.2004.04.007.

[59] Jeon, J., Victor, M., Adler, S. P., Arwady, A., Demmler, G., Fowler, K., Goldfarb, J., Keyserling, H., Massoudi, M., Richards, K., Staras, S. A., and Cannon, M. J. (2006). Knowledge and awareness of congenital cytomegalovirus among women. *Infectious Diseases in Obstetrics and Gynecology*, 2006: 80383. doi: 10.1155/IDOG/2006/80383.

[60] Fowler, K. B. and Pass, R. F. 2006. Risk factors for congenital cytomegalovirus infection in the offspring of young women: exposure to young children and recent onset of sexual activity. *Pediatrics*, 118 (2): e286–292. doi: 10.1542/peds.2005-1142.

Chapter 9

Loop-mediated Isothermal Amplification: An Overview and Its Applications

Ranjita Chatterjee[1]
Sourabh Sulabh[1]
Ritu Varshney[2]
and Rajat Varshney[3,*]
[1]Department of Animal Science, Kazi Nazrul University, Asansol, West Bengal, India
[2]Biological Engineering, Indian Institute of Technology Gandhinagar, Gujrat, India
[3]Department of Veterinary Microbiology, Faculty of Veterinary and Animal Sciences, RGSC, Banaras Hindu University, Mirzapur, Uttar Pradesh, India

Abstract

Loop-mediated isothermal amplification (LAMP) is a valuable method of nucleic acid amplification which has made revolutionary changes in the detection of genetic traits, genetic malfunctions, and the presence of infectious diseases. LAMP amplifies DNA with very high specificity, efficiency, and rapidity under isothermal conditions by using a DNA polymerase with high displacement strand activity and a set of specifically designed primers to amplify specified DNA strands. After its invention, LAMP was modified over a period, which made use of the combination of this method with other molecular approaches, such as the reverse transcription technique and multiplex amplification method, to distinguish infectious diseases in humans, livestock, and plants. Till now, several types of LAMP detection techniques are available, including colorimetric and fluorescent detection, real-time monitoring using a turbidity meter, and detection using the lateral flow device. Moreover,

* Corresponding Author's Email:varshney.rajat1@gmail.com.

In: Advances in Biology. Volume 4
Editor: Charles D. Grant
ISBN: 979-8-88697-999-2

the LAMP technique has also been commercialized in the form of lyophilized LAMP reagent kits and primer sets to identify and mark pathogenic microorganisms. The clinical utility of LAMP is quite promising as a point-of care-test in the resource-limited laboratory and fields as this isothermal amplification technique precludes the need for the thermo-cycler.

Keywords: LAMP, DNA detection method, diagnosis, infectious diseases, point-of-care test

Introduction

Precise, rapid, and sensitive diagnostic methods are required for disease containment and early management of outbreaks (Varshney et al., 2017). Early disease diagnosis and prompt institution of pathogen-specific directed therapy limit infectious diseases and coordinate antimicrobial stewardship by preventing the evolution of drug-resistant microorganisms (Agrawal et al., 2020; Varshney et al., 2020). Moreover, 'test and 'segregation' or 'test and 'slaughter' policies in animals are usually followed in the absence of reliable vaccines and therapy. The shortcomings of conventional diagnostic methods viz. unculturable or slow-growing microorganisms, stringent cultivation requirements, time-consuming intricacies cultivation procedures, strain-wise variations in biochemical tests, anergy or overactivity-related constraints in intradermal testing, etc., set inclinations towards molecular tests especially nucleic acid amplification techniques (NATs). NATs are currently not only used as a biological technique but are also being adapted in various scientific fields, including forensic science, agriculture, clinical diagnosis, epidemiology, etc., (Moore, 2005).

NAT in the laboratory starts with the visitation of polymerase chain reaction (PCR), which involves three prime cyclic reactions [i.e., denaturation, annealing, and elongation] at different temperatures. The PCR assay is highly sensitive, specific, and faster than the time-consuming conventional microbiological approaches, but the requirement of sophisticated equipment puts it beyond the budgetary capacity of most field diagnostic centers or resource-limited centers, especially in the veterinary sector. Therefore, a fundamental problem in current veterinary research and animal health care is developing novel, easy-to-perform, and affordable pen-site diagnostic tests as point-of-care tests (POCT). Now isothermal methods are trending as the reaction is performed at constant temperature; an energy-intensive thermocycler is not needed (Varshney et al, 2017).

Over time various variants of PCR, real-time PCR (Panno et al., 2019), ligase chain reaction, nucleic acid sequence-based amplification (Mugasa et al., 2009), molecular hybridization (Puchades et al., 2017), flow-through hybridization (Wetmur 1991), SDA (Walker et al., 1992), rolling circle amplification (RCA), single primer isothermal amplification, and circular helicase dependent amplification (Gill and Ghaemi 2008) have been developed for diagnostic tests. However, the requirement of a well-equipped laboratory, costly reagents, and highly skilled personnel limit their applicability. Ultimately, Notomi et al., (2000) invented a novel technique that amplifies targeted nucleic acid sequence under isothermal conditions within an hour. This technique is termed loop-mediated isothermal amplification (LAMP).

Principles of LAMP

The LAMP reaction involves an initial step, a cycling amplification step, and an elongation step (Mori et al., 2009), which is executed [i.e., the optimum temperature for Bst polymerase activity] at 60–65 °C (Chander et al., 2014). The amplification efficiency of the LAMP remains exceptionally high as time loss for the thermal change is not occur. The denaturation is not required in the LAMP technique because of the strand displacement activity of Bst DNA polymerase (source-*Geobacillus stearothermophilus*) or Bsm polymerase (source-*Bacillus smithii*). Moreover, both enzymes can catalyze 5'-3' DNA polymerization but lack 5'-3' exonuclease activity. The use of betaine in LAMP is frequent to lower the melting temperature, facilitate the separation of DNA templates and increase the specificity. LAMP reactions require nearly seven times more dNTPs than conventional PCR due to the generation of multiple DNA products of different sizes.

Four (or six) different primers used in the LAMP assay recognize six (or eight) different regions on the target gene making it immensely specific. This primer set comprises two outer (F3 and B3) primers, two inner primers [forward inner primer (FIP) and backward inner primer (BIP)] along with loop primers [loop forward (LF) and loop backward (LB)]. Inner primers (FIP or BIP) include two distinctive sequences that diagnose a sense (F2 or B2) and an antisense sequence (F1c or B1c) of the target DNA; meanwhile, the outer primers (F3 or B3) identify only one external sequence of the target DNA (Notomi et al., 2000).

General and Specific Considerations for Primers Design

In various nucleic acid amplification techniques, designing primers and/or probes is one of the most vital points that have a direct impact on the success and quality of the results (Abd-Elsalam K.A. 2003). While construction of the LAMP primers uses the preference of eight different regions of the desired nucleic acid sequence.

Special attention is paid to avert the formation of the dimers among primers, especially about 40 nucleotide long FIP and BIP primers. FIP comprises the F2 sequence (at its 3′ end) complementary to the F2c region and the exact sequence as the F1c region at its 5′ end. BIP comprises the B2 sequence (at its 3′ end) complementary to the B2c region, and the exact sequence as B1c region at its 5′ end. LF primer is designed with the help of the complementary strand corresponding to the area between F1 and F2, meanwhile, LB primer is designed with the use of the complementary strand corresponding to the region between B1 and B2 (Notomi et al., 2000).

Key Factors for Primer Design

Primer length, Guanine and Cytosine (GC) content (optimal GC content range between 40% and 60%), no-specific binding, stability at the 3' and 5' ends of each primer, primer secondary structures, and melting temperature (T_m) are some of the important factors that need to be considered when designing primers. It is also vital to consider the distance between primers, primer secondary structures (self-dimers and cross-dimers), and T_m balance since numerous primers simultaneously anneal in various target regions.

LAMP Primer Design Software

Due to the complexity of primer design, several software programs have been created and are available online. Primer Explorer V4 (PE4), an accessible primer design tool provided by Eiken Chemical Co. (Primer Explorer), is frequently used. Additional software has been developed to implement and improve certain features; however, most of them are expensive and require a license. LAVA (LAMP Assay Versatile Analysis) (Torres et al., 2011) and LAMP Designer (Optigene, Horsham, UK) are two of them. These tools allow for avoiding cross homologies during primer production and facilitate primer design via multiple sequence alignment.

Steps in LAMP Methods

Initial Step

As double-stranded DNA remains in dynamic equilibrium condition around 65 °C, the F2 region of FIP may anneal to the complementary sequence (F2c) of target DNA, then initiates DNA synthesis using the Bst or Bsm DNA polymerase, displacing and releasing a single-stranded DNA. Also, a DNA strand is synthesized which is complementary to the template DNA (Figure 1). This complementary DNA is synthesized starting from the 3' end of the F2 region of the FIP. The F3 primer anneals to the F3c region of target DNA outside of the FIP sequence and starts F3 primed DNA strand syntheses from the template DNA strand, releasing the FIP-linked complementary strand. Because of the complementary F1c and F1 regions, the split single strand develops a self-hybridizing loop (stem-loop structure) at the 5' end. It serves as a template for BIP-initiated DNA synthesis as well as subsequent B3-primed strand displacement DNA synthesis. The B2 region of BIP anneal to the B2c sequence of the template strand, and BIP primed complementary DNA synthesis reverts the template stem-loop structure into a linear structure. The B3 primer anneals to the target DNA's B3c region outside of the BIP sequence, and starts B3-primed DNA strand synthesis from the template DNA strand, releasing the BIP-linked complementary strand. This released BIP-linked complementary strand forms stem loops at each end and exhibits a dumbbell-like structure (Figure 1). This structure serves as the starting structure for the LAMP amplification cycle.

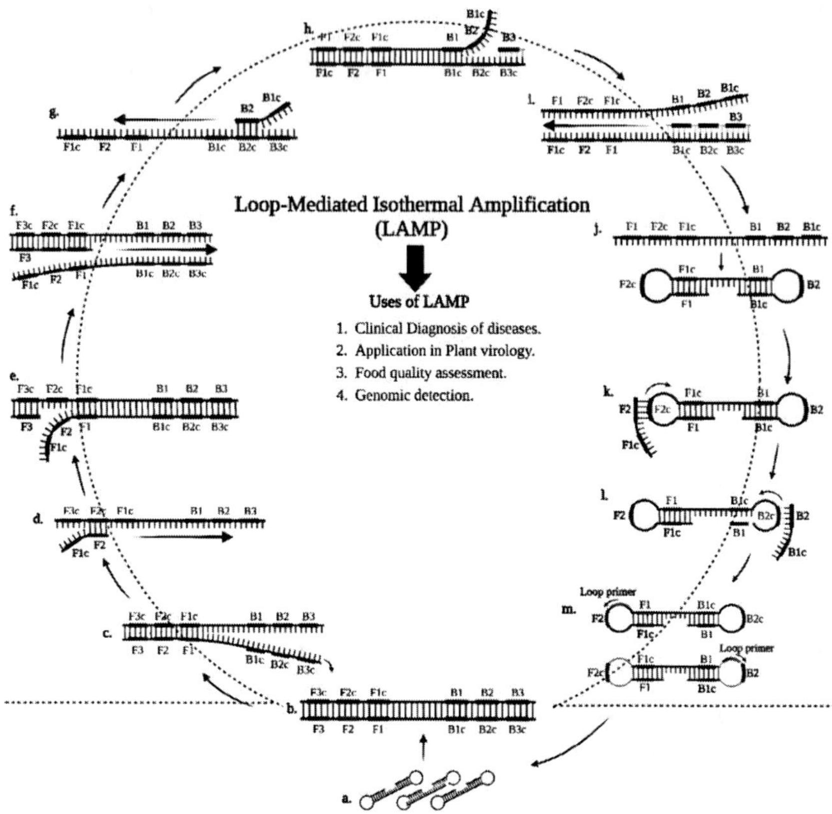

Figure 1. Schematic representation of LAMP reaction and its principle, along with some applications.

(a) Target DNA, (b) Target sequence, (c) DNA denaturation, (d) Forward Inner Primer (FIP) that is F1c-F2 binds to F2c region for complementary polymerization, (e) F3 primer binds to F3c region, (f) F3 performs strand displacement for F1c-F2 strand to initiate a new polymerization cycle, (g) Backword Inner Primer (BIP) that is B2-B1c binds to the displaced strand and initiates polymerization, (h) B3 binds to the B3c region, (i) B3 performs strand displacement for the B2-B1c strand to initiate a new polymerization cycle, (j) Through complementary reaction, B2-B1c resultant amplicon forms a "Dumbbell" structure, (k) A complementary strand synthesized as an F1c-F2 primer binds to the "dumbbell" region, (l) B2-B1cprimer binds to the "dumbbell" region to synthesize new strand for the progression of the LAMP cycle, (m) To enhance the shape complexity of the amplicons additional loop primers can be added. At the end multiple copies of the target sequence are generated) (Notomi et al., 2000).

Cycling Amplification and Elongation Step

FIP hybridizes to the loop structure of the strand created during an initial step. The two products of self-primmed strand displacement DNA synthesis are a complementary strand and a second strand with a double-elongated stem that is the same length as the first and the loop on the opposite side. The ensuing elongation and recycling stages utilize both strands as templates for the reverse-primed strand displacement synthesis. As a result, the target sequence can be amplified three times during each half of the cycle (Notomi et al., 2000). Due to the Bst DNA polymerase's strong displacement activity, a significant amount of DNA with high molecular weight is produced quickly (Hadersdorfer et al., 2011). As a result, target DNA can be amplified up to 10^9 copies in less time. Ultimately, several looped stem-loop DNAs with varying lengths are created to resemble cauliflower-like formations (Mori et al., 2006).

LAMP Acceleration

Two additional primers (LF and LB) could be added to speed up the reaction. These primers hybridize to the stem strands except for the loops that hybridize to internal primers and primer displacement DNA synthesis (Nagamine et al., 2002). Currently, six primers are effectively applied to various diagnostic protocols to avail superior specificity and sensitivity (Varga et al., 2006). A variant application of this technology was based on the capability of the Bst polymerase to tolerate different types of inhibitors (Francois et al., 2011). Unlike the pre-existing methods, LAMP takes much less time and with minimal components, making it more acceptable at the grass root level.

Monitoring of LAMP Products

The ladder-like pattern on agarose gel electrophoresis (AGE) indicates positive LAMP results, which is due to the generation of many DNA bands of different sizes. The LAMP amplified product can be detected by turbidity due to the formation of white magnesium pyrophosphate in solution or photometry or with the addition of colorimetric indicators viz. SYBR green, Pico green dye, calcein, hydroxy naphthol blue (HNB), etc. In contrast, turbidity is not observed in PCR due to the hydrolysis of pyrophosphate at the high temperatures used in the denaturation step. The other various methods viz. fluorescence methods, electrochemical sensors/chips, electrochemical biosensors, lateral flow dipstick, Enzyme-linked immunosorbent assay, NanoAu probe, Field effect sensor, Surface plasmon resonance sensor, the absorbance of amplicons, Giant magnetoresistive sensor and bioluminescence can also be used for monitoring LAMP products.

Applications of LAMP

LAMP as POCT needs to be improved and integrated with simple genetic tests (Mori et al., 2013). LAMP technology has been used to detect various human, plant as well as animal pathogens, and monitor, and quarantine control of diseases. In addition, LAMP is used to detect SNPs, transgenic products, cancer, genetically modified organisms, and embryo sex (Fu et al., 2011). LAMP can be performed with a regular block heater or water bath without a thermocycler. LaBarre and colleagues (2011) have created an electricity-free heater based on exothermic chemical reactions and engineered phase change materials that can be used for any isothermal nucleic acid amplification assays, such as LAMP. Various "versions" of LAMP have been created in recent years.

Multiplex LAMP (mLAMP)

Multiplex LAMP (mLAMP), like multiplex PCR, can detect multiple infections simultaneously. In 2007, Iseki and colleagues developed the mLAMP method by combining two sets of LAMP primers with a restriction enzyme cleavage site in two pairs of species-specific primers. Consequently, this approach enables the simultaneous detection of both pathogens. mLAMP has undergone numerous modifications and enhancements; nonetheless, they require more advanced instruments. Real-time mLAMP allows the real-time detection of 1-4 target sequences in a single LAMP reaction tube using a conventional real-time fluorimeter and typical LAMP primers that contain a quencher-fluorophore duplex region that, upon strand separation, results in a gain of fluorescence signal. (Tanner et al., 2012).

Micro LAMP (μLAMP)

The μLAMP assay requires a modest sample volume (0.4 L). It can be integrated onto a microfluidic chip for measurement by an optic sensor or for readout by the naked eye (insoluble by-product pyrophosphate). The system can be changed to work with a real-time quantitative LAMP by integrating optical fibers inside the device (Fang et al., 2010).

Digital LAMP (dLAMP)

In 2012, the first on-chip LAMP in a digital format was demonstrated using a digital technique created using a sample self-digitization chip. The sample is pipetted directly into the chip and a basic pump head. The constant air pressure cause sample discretization for the subsequent isothermal amplification, allowing this system to detect relative changes in template concentration as well as absolute quantification of template numbers. Within a 70-minute incubation period at 60-65 °C, dLAMP enables accurate absolute and relative DNA quantification. Lin and co-workers (2019) reported a new dLAMP method that uses a commercial track-etched polycarbonate membrane instead of a chip, without the need for specialized equipment, to simplify and lower effective costs. This membrane system could be used for digital quantification that is flexible and POCT.

LAMP Combined with Lateral Flow Assay (LFA)

Biotin-labeling and fluorescein isothiocyanate (FITC)-labeling of LAMP products can be done during or after amplification. Biotin labeling of the LAMP product during amplification and a FITC DNA probe is added to hybridize the biotinylated LAMP product after the LAMP reaction. The FITC and biotin-labeled LAMP product is loaded on the sample loading area on the lateral flow test strip. Only when anti-FITC-Ab-coated gold nanoparticles bind with the labeled LAMP product does the test line emerge on the strip, while the control line serves as an inherent control for the validity of LFA (Agarwal et al., 2022). This combination is a portable, easy-to-perform pen-site test and requires no specialized equipment, so it is clinical-friendly.

Immunocapture-RT-LAMP (IC-RT-LAMP)

Selvaraj and colleagues (2019) enhanced an RT-LAMP assay for citrus tristeza virus (CTV) detection by first capturing CTV virions from citrus crude leaf sap using CTV-IgG. This facilitated quick on-site detection. This process enables the RNA extraction step to be skipped, saving time and money while maintaining high sensibility and robustness. In practice, inexperienced workers can complete IC-RT-LAMP in around an hour and a half.

Electric LAMP (eLAMP)

The eLAMP is a technological tool designed to improve the efficiency of testing potential LAMP primers on target gene sequences using a graphical interface that electronically simulates the LAMP assay, a Practical Extraction and Report Language (PERL) script; it may be helpful to ascertain whether it is feasible to use an existing primer to amplify recently discovered sequence variants. (Salinas and Little, 2012).

Lab-on-a-Disc (LoaD) LAMP

This fully integrated, portable "Lab-on-a-Disc" (LoaD) system enables automated detection of plant pathogens through quantitative LAMP amplification. The platform is made up of a disposable cartridge disc that rotates around its axis to create a centrifugal field. The resulting pressure head pumps the sample outward from the disc's center to its edge. In more detail, digital pulse-actuated dissolvable-film valves directly process, purify, and then combine samples with LAMP reagents. Heating and fluorescent detection modules are built into this unique instrument. This all-in-one system has a lot of promise for decentralized and on-site applications like cancer diagnostics and food quality monitoring. Although this system was created to detect Botrytis cinerea, it is easily adaptable to detecting viruses and other pathogens (Kinahan et al., 2018).

Lyophilized LAMP

To simplify and speed up the LAMP process, this version of LAMP combines all the reagents into a single mixture known as the lyophilized LAMP mix, sometimes in conjunction with a closed amplification and detection system. The user only needs to add nuclease-free water (NFW) and a sample or DNA/RNA template to the lyophilized mix before incubating at the desired temperature. Commercially available lyophilized LAMP kits are widely used for rapid disease diagnosis, enabling their use in the field and areas with limited resources (Masilamani et al., 2013; Abdullahi et al., 2015). These kits are typically used in conjunction with portable thermal cyclers, which enable wireless connectivity with portable devices (like smartphones or tablets) for real time data viewing. This system can be used to quickly diagnose the potential presence of a plant pathogen in the agricultural and phytosanitary fields, especially when combined with decision support tools (Davino et al., 2017).

FRET-based Assimilating Probe for Sequence-Specific Real-time Monitoring of LAMP

A fluorescence resonance energy transfer (FRET) based probe is the foundation of this method. Using field samples, this method enables the precise detection and target quantification of LAMP products (Kubota et al., 2011). For real-time sequence-specific monitoring in this LAMP version, a pair of labeled oligonucleotide probes (assimilating probes) is used following the FRET method. The probe pair works like a "molecular zipper," separating a quenching strand from a partially complementary fluorescent strand during the DNA synthesis process. A molecular zipper is a fluorescent probe design that is utilized when target DNA is quantified using real-time monitoring ramification amplification (RAM) or RCA reactions. This probe has very low background fluorescence because of the strong connection between the two strands. When it is introduced into RAM products, its double-strand area is opened by displacement, and its fluorophore subsequently emits a fluorescent signal. (Yi et al., 2006). Utilizing this technology significantly minimizes any potential error resulting from the visual detection of LAMP products. Additionally, it increases the precision of pathogen detection and simultaneously permits their quantification.

Advantages and Drawbacks of LAMP Assay

LAMP usually exhibits more sensitivity and specificity in contrast to PCR. LAMP precludes the requirement of a thermocycler, and post-amplification processing as LAMP can be operated at a constant temperature in a water bath, and LAMP products can be easily visualized using SYBR Green I, EtBr, HNB, or Calcein with the human eye. The activity of Bst polymerase remains intact even in the presence of inhibitory substances, which in turn justifies the robustness of the LAMP. Therefore, LAMP can be performed by using direct crude extraction rather than total RNA or DNA extraction, which would reduce the processing time, enable the analysis of multiple samples at once, and significantly lower the cost of a single analysis. Amplification of RNA sequences can also be accomplished in a LAMP assay with high efficiency when combined with reverse transcription.

The proper primer design is the primary constraint of the LAMP assay. Consequently, multiplexing techniques for LAMP are less advanced than those for PCR. LAMP is also not appropriate for cloning because the results are a mixture of stem-loop DNAs and a structure resembling a cauliflower. Because of its high sensitivity, the LAMP method is susceptible to contamination. However, the problem of carryover contamination in LAMP can be resolved by pre-incubating the LAMP mixture with uracil-N-glycosylase or using DNA binding dyes or metal ion indicators in the LAMP mixture. The tube does not need to be opened in case of using DNA binding dyes/metal ion indicators, and the detection is carried out using a single "closed-tube" technology. Finally, master mix preparation should be made on ice in less than 30 minutes to prevent contaminations.

Conclusion

LAMP is considered to be a revolution in the field of detection of gene amplification. It works as a quick and simple diagnostic tool for rapidly detecting and identifying several pathogens. LAMP is simple to use and readily adaptable to all field conditions and locations, and it offers all the capabilities required for real-time assays, particularly in high sensitivity. LAMP is also a beneficial instrument for developing countries due to its ease of use and lack of requirement for sophisticated equipment or expertise to operate it.

References

Abd-Elsalam K. A. Bioinformatic tools and guidelines for PCR primer design. *Afr. J. Biotechnol.* 2003;2:91–95.

Abdullahi U. F., Naim R., Taib W. R. W., Saleh A., Muazu A., Aliyu S., Baig A.A. Loop-mediated isothermal amplification (LAMP), an innovation in gene amplification: Bridging the gap in molecular diagnostics; a review. *Indian J. Sci. Technol.* 2015;8:1. doi: 10.17485/ijst/2015/v8i17/55767.

Agarwal, S., Warmt, C., Henkel, J., Livia Schrick, Andreas Nitsche, Frank F Bier. Lateral flow–based nucleic acid detection of SARS-CoV-2 using enzymatic incorporation of biotin-labeleddUTP for POCT use. *Anal. Bioanal. Chem.* 414, 3177–3186 (2022). https://doi.org/10.1007/s00216-022-03880-4.

Agrawal A, Varshney R, Gattani A, Kirthika P, Khan M H, Singh R, Kodape S, Patel SK, Singh P.Gold nanoparticle based immunochromatographic biosensor for rapid diagnosis of Mycobacterium avium subspecies paratuberculosis infection using recombinant protein. *J. Microbiol. Methods.* 2020 Oct;177:106024. doi: 10.1016/j.mimet.2020.106024. Epub 2020 Aug 12. PMID: 32795639.

Chander Y., Koelbl J., Puckett J., Moser M. J., Klingele A. J., Liles M. R., Carrias A., Mead D. A., Schoenfeld T.W. A novel thermostable polymerase for RNA and DNA loop-mediated isothermal amplification (LAMP) *Front. Microbiol.* 2014;5:395. doi: 10.3389/fmicb.2014.00395.

Davino S., Panno S., Arrigo M., La Rocca M., Caruso A.G., Bosco G.L. Planthology: An application system for plant diseases management. *Chem. Eng. Trans.* 2017;58:619–624.

Fang X., Liu Y., Kong J., Jiang X. Loop-mediated isothermal amplification integrated on microfluidic chips for point-of-care quantitative detection of pathogens. *Anal. Chem.* 2010;82:3002–3006. doi: 10.1021/ac1000652.

Francois P., Tangomo M., Hibbs J., Bonetti E. J., Boehme C. C., Notomi T., Perkins M. D., Schrenzel J. Robustness of a loop-mediated isothermal amplification reaction for diagnostic applications. *FEMS Immunol. Med. Microbiol.* 2011;62:41–48. doi: 10.1111/j.1574-695X.2011.00785.x.

Fu S., Qu G., Guo S., Ma L., Zhang N., Zhang S., Zhang S., Gao S., Shen Z. Applications of loop-mediated isothermal DNA amplification. *Appl. Biochem. Biotechnol.* 2011;163:845–850. doi: 10.1007/s12010-010-9088-8.

Gill P., Ghaemi A. Nucleic acid isothermal amplification technologies—A review. *Nucleosides Nucleotides Nucleic Acids.* 2008; 27:224–243. doi: 10.1080/15257770701845204.

Hadersdorfer J., Neumüller M., Treutter D., Fischer T.C. Fast and reliable detection of Plum pox virus in woody host plants using the Blue LAMP protocol. *Ann. Appl. Biol.* 2011;159:456–466. doi: 10.1111/j.1744-7348.2011.00510.x.

Iseki H., Alhassan A., Ohta N., Thekisoe O.M., Yokoyama N., Inoue N., Nambota A., Yasuda J., Igarashi I. Development of a multiplex loop-mediated isothermal amplification (mLAMP) method for the simultaneous detection of bovine Babesia parasites. *J. Microbiol. Methods.* 2007;71:281–287. doi: 10.1016/j.mimet.2007.09.019.

Kinahan D. J., Julius L. A., Schoen C., Dreo T., Ducrée J. Automated DNA purification and multiplexed lamp assay preparation on a centrifugal microfluidic "Lab-on-a-Disc" platform; *Proceedings of the 2018 IEEE Micro Electro Mechanical Systems (MEMS);* Belfast, UK. 21–25 January 2018; pp. 1134–1137.

Kubota R., Alvarez A. M., Su W. W., Jenkins D. M. FRET-based assimilating probe for sequence specific real-time monitoring of loop-mediated isothermal amplification (LAMP) *Biol. Eng. Trans.* 2011;4:81–100. doi: 10.13031/2013.38509.

LaBarre P., Hawkins K. R., Gerlach J., Wilmoth J., Beddoe A., Singleton J., Boyle D., Weigl B. A simple, inexpensive device for nucleic acid amplification without electricity—Toward instrument-free molecular diagnostics in low-resource settings. *PLoS ONE.* 2011;6:e19738. doi: 10.1371/journal.pone.0019738.

Lin X., Huang X., Urmann K., Xie X., Hoffmann M.R. Digital Loop-Mediated Isothermal Amplification on a Commercial Membrane. *ACS Sens.* 2019;4:242–249. doi: 10.1021/acssensors.8b01419.

Masilamani S. M., Sasitharan D., Manohar P. W. Associated technologies ensures complete loop mediated isothermal amplification platform for pathogen diagnosis. *Afr. J. Biotechnol.* 2013:12. doi: 10.5897/AJB2013.12399.

Moore, P., 2005. *Nature* 435 (7039), 235-238.

Mori Y., Hirano T., Notomi T. Sequence specific visual detection of LAMP reactions by addition of cationic polymers. *BMC Biotechnol.* 2006;6:3. doi: 10.1186/1472-6750-6-3.

Mori Y., Kanda H., Notomi T. Loop-mediated isothermal amplification (LAMP): Recent progress in research and development. *J. Infect. Chemother.* 2013;19:404–411. doi: 10.1007/s10156-013-0590-0.

Mori Y., Notomi T. Loop-mediated isothermal amplification (LAMP): A rapid, accurate, and cost-effective diagnostic method for infectious diseases. *J. Infect. Chemother.* 2009;15:62–69. doi: 10.1007/s10156-009-0669-9.

Mugasa C. M., Laurent T., Schoone G.J., Kager P. A., Lubega G. W., Schallig H. D. Nucleic acid sequence-based amplification with oligochromatography for detection of Trypanosoma brucei in clinical samples. *J. Clin. Microbiol.* 2009;47:630–635. doi: 10.1128/JCM.01430-08.

Nagamine K., Hase T., Notomi T. Accelerated reaction by loop-mediated isothermal amplification using loop primers. *Mol. Cell. Probes.* 2002;16:223–229. doi: 10.1006/mcpr.2002.0415.

Notomi, T., Okayama, H., Masubuchi, H., Yonekawa, T., Watanabe, K., Amino, N. and Hase, T. (2000) Loop-mediated isothermal amplification of DNA. *Nucleic Acids Res.* 28, E63.

Panno S., Ruiz-Ruiz S., Caruso A. G., Alfaro-Fernandez A., San Ambrosio M. I. F., Davino S. Real-time reverse transcription polymerase chain reaction development for rapid detection of Tomato brown rugose fruit virus and comparison with other techniques. *Peer J.* 2019;7:e7928. doi: 10.7717/peerj.7928.

PrimerExplorer. [(accessed on 3 February 2020)]; Available online: https://primerexplorer.jp/e/.

Puchades A.V., Carpino C., Alfaro-Fernandez A., Font-San-Ambrosio M.I., Davino S., Guerri J., Rubio L., Galipienso L. Detection of Southern tomato virus by molecular hybridisation. *Ann. Appl. Biol.* 2017;171:172–178. doi: 10.1111/aab.12367.

Rigano L.A., Malamud F., Orce I.G., Filippone M.P., Marano M.R., Do Amaral A.M., Castagnaro A. P., Vojnov A. A. Rapid and sensitive detection of Candidatus *Liberibacter asiaticus* by loop mediated isothermal amplification combined with a lateral flow dipstick. *BMC Microbiol.* 2014;14:86. doi: 10.1186/1471-2180-14-86.

Salinas N. R., Little D. P. Electric LAMP: Virtual loop-mediated isothermal AMPlification. *ISRN Bioinform.* 2012;2012 doi: 10.5402/2012/696758.

Selvaraj V., Maheshwari Y., Hajeri S., Yokomi R. A rapid detection tool for VT isolates of *Citrus tristeza virus* by immunocapture-reverse transcriptase loop-mediated isothermal amplification assay. *PLoS ONE.* 2019;14:e0222170. doi: 10.1371/ journal. pone.0222170.

Torres C., Vitalis E.A., Baker B. R., Gardner S. N., Torres M. W., Dzenitis J. M. LAVA: An Open-Source approach to designing LAMP (Loop-Mediated Isothermal Amplification) DNA signatures. *BMC Bioinform.* 2011;12:240–247. doi: 10.1186/ 1471-2105-12-240.

Varga A., James D. Use of reverse transcription loop-mediated isothermal amplification for the detection of *Plum pox* virus. *J. Virol. Methods.* 2006;138:184–190. doi: 10.1016/j.jviromet.2006.08.014.

Varshney R, Varshney R, Chaturvedi V K, Rawat M, Saminathan M, Singh V, Singh R, Sahoo M, Gupta P K. Development of novel iron-regulated *Pasteurella multocida* B: 2 bacterin and refinement of vaccine quality in terms of minimum variation in particle size and distribution vis-a-vis critical level of iron in media. *Microb Pathog.* 2020 Oct;147:104375. doi: 10.1016/j.micpath.2020.104375. *Epub 2020* Jul 15. PMID: 32679244.

Varshney, R., Singh, A. & Sharma, A. Development of Loop-Mediated Isothermal Amplification Assay Based on *6b* Gene Sequence for Rapid Detection of *Pasteurella multocida* B:2. *Proc. Natl. Acad. Sci., India, Sect. B Biol. Sci.* 87, 1377–1384 (2017). https://doi.org/10.1007/s40011-016-0713-3.

Walker G. T., Little M. C., Nadeau J. G., Shank D. D. Isothermal in vitro amplification of DNA by a restriction enzyme/DNA polymerase system. *Proc. Natl. Acad. Sci.* USA. 1992;89:392–396. doi: 10.1073/pnas.89.1.392.

Wetmur J. G. DNA probes: Applications of the principles of nucleic acid hybridization. *Crit. Rev. Biochem. Mol. Biol.* 1991;26:227–259. doi:10.3109/ 10409239109114069.

Yi J. Z., Zhang W. D., Zhang D.Y. Molecular Zipper: A fluorescent probe for realtime isothermal DNA amplification. *Nucleic Acids Res.* 2006;34:e81. doi:10.1093/ nar/gkl261.

Chapter 10

Short-term Liquid Nitrogen Storage of *Leucaena leucocephala* Seeds and Phenotypic, Phytochemical and Biochemical Analyses in 21-Day-Old Seedlings

Sara Thamires Dias da Fonseca[1], MS
Mairon César Coimbra[2], PhD
and Ana Hortência Fonsêca Castro[2,*], PhD
[1]Universidade de São Paulo, Faculdade de Ciências Farmacêuticas, Ribeirão Preto, São Paulo, Brazil
[2]Universidade Federal de São João del-Rei, Divinópolis, Minas Gerais, Brazil

Abstract

Leucaena leucocephala belongs to the Fabaceae family and has nutritive value, potential uses, such as anthelmintic, anti-cancer, anti-diabetic, anti-viral, and immunostimulant treatments, other non-medicinal uses, such as forage, timber, firewood, and raw material for pulp and paper in food and pharmaceutical industries, and nutritional uses. However, the seeds present dormancy that can reduce the rate of germination to less than 50% and increase weed infestation. In this work, we studied the effect of short-term liquid nitrogen storage of seeds on the germination, phenotypic characteristics, and phenolics compound, soluble sugar and protein contents in *L. leucocephala*. In the cryostorage assay, seeds were immersed in liquid nitrogen (LN) for 7, 14 and 21 days and thawed for 30, 60, 90 and 120 seconds. After freezing and thawing, the seeds were placed on basal Murashige and Skoog medium for 21 days, and then

* Corresponding Author's Email: acastro@ufsj.edu.br.

In: Advances in Biology. Volume 4
Editor: Charles D. Grant
ISBN: 979-8-88697-999-2

phenotypic, phytochemical and biochemical analyses were performed. Depending on the treatment, the seeds had reduced germination percentages by up to 62.5%. The phenotypic analyses revealed that the longer the exposure time of the seeds to LN, the greater the fresh and dry weights and shoot and root lengths of the seedlings. In general, the phenolic compound content was higher in the shoots, while the levels of sugars and total proteins were higher in the roots. Therefore, further studies are required to clarify the mechanisms involved in the observed results.

Keywords: cryostorage, phenotype variation, biochemical changes, phenolic compounds

Introduction

Native to Central America, *Leucaena leucocephala* (Lam.) de Wit, also known as "Leucena," belongs to the Fabaceae family and has nutritive values and potential uses. It is a perennial, thornless tropical plant with a height of around 8 m that can be found in many tropical and subtropical regions of the world (Soares et al., 2015). All of its plant parts have secondary metabolites, and the seeds have a high nutritional value because of their high protein content (Verma et al., 2018). *Leucaena leucocephala* has medicinal properties, such as anti-proliferative, cancer chemo-protective, anthelmintic, anti-bacterial, anti-viral, anti-thrombotic, immunostimulant, anti-diabetic and anti-inflammatory effects (Burawat et al., 2016; Soares et al., 2015). There are many reports about the presence of phenolic compounds and other compounds with antioxidant potential in extracts from several plant organs. In addition to medicinal uses, *L. leucocephala* is also used for nutrition, forage, timber, firewood, and raw material for pulp and paper (Mittal, Mattu and Kaur, 2016).

Despite its wide distribution and use, *L. leucocephala* seeds exhibit dormancy caused by the presence of a resistant and impermeable tegument that prevents the entry of water and gas exchange and does not allow germination (Alves et al., 2000). Thus, to produce *L. leucocephala* seedlings, it is necessary to break natural dormancy. The germination rate of *L. leucocephala* seeds without breaking dormancy is less than 50% and causes slow and irregular emergence, in addition to favoring weed infestation (Teles et al., 2000).

Due to the difficulty of propagating *L. leucocephala* and preserving it from unsustainable exploitation, biotechnology can be used as an important tool for the characterization, conservation and use of the species. *In vitro* culture techniques allow the development of studies that establish alternative forms for the conservation and improvement of genetic material, in addition to contributing to the optimization of the production of plant compounds of pharmaceutical interest (Pêgo, Paiva and Paiva, 2013). Studies of Fabaceae species have shown that seedling production from in vitro seed germination is economically feasible (Nascimento, Frando and Frassetto, 2007), and integument dormancy can be overcome by the mechanical excision of seeds (Alves et al., 2000).

Although *in vitro* cultures provide plant propagation materials independent of the time of year, it is necessary to use specific techniques for the conservation of long-term biotechnology products. Cryopreservation is a technique used for the conservation of several plant species by storage of seeds, shoot apices, nodal segments and cells suspended in liquid nitrogen (LN) for long periods (Chen et al., 2011). In ultra-low temperatures, the metabolic processes are interrupted, which allows the storage of a large amount of material in a small space for an unlimited time without any changes in the stored material (Engelmann, 2011).

Seeds are one of the most used sources of germplasm for the *ex situ* preservation of genetic resources for agriculture, horticulture, forestry and biodiversity conservation (Li and Prichard, 2009).

This study aimed to assess the phenotypic, biochemical and phytochemical profiles of *L. leucocephala* seedlings produced from cryostored seeds in LN. Thus, this work presents itself as an informational resource for biotechnological strategies for the conservation and propagation of this species.

Matherial and Methods

Plant Material

Leucaena leucocephala seeds were collected in Divinópolis, Midwest Minas Gerais State, Brazil (20°10'45.9" S and 44°55'07.2" W GRW) (SISBIO no. 24542-7). Fertile samples were collected, and vouchers were identified by Andreia Fonseca Silva of the PAMG Herbarium (PAMG 58034) at the Agricultural Research Company of Minas Gerais (EPAMIG). This work was registered in the SisGen Platform (Register A88527D), according to Brazilian

Biodiversity Law 13.123/2015. Seeds were treated for 5 minutes with Captan in the proportion of 1 g/kg of seeds, according to Pires, Bragantini and Costa (2004) and stored in a cold chamber until use.

Freezing Tolerance

Seeds were placed in cryotubes and cryostored in liquid nitrogen (LN) (−196°C) for 7, 14 and 21 days. After this, the cryotubes were removed from the LN and thawed directly in a water bath for 30, 60, 90 and 120 seconds at 40 ± 1°C. The cryotubes were taken to the laminar flow hood where they were opened. The seeds and embryos received aseptic treatment and were inoculated. The test was performed in triplicate.

In Vitro Germination and Phenotypic Analyses of Seedlings

After thawing, the seeds were disinfected in 2% NaClO (commercial sodium hypochlorite) for 10 min and washed five times with sterile distilled water. After disinfestation, the seeds were excised and placed on Murashige and Skoog (MS) medium (Murashige and Skoog, 1962) supplemented with 30 g L^{-1} sucrose and solidified with 7 g L^{-1} agar. The pH was adjusted to 5.7 ± 0.1. The seeds were kept in a growth chamber at 27 ± 2°C under the presence of a light (16:8 h light/dark regime, with a light intensity of 40 μmol m^{-2} s^{-1}). A completely randomized design was adopted, consisting of 12 treatments with 25 seeds each and 3 repetitions. The control was represented by non-cryostored seeds. After 21 days, the following characteristics were evaluated: a) for seeds: the germination speed index (GSI) (Maguire, 1962), seed germination, contamination and oxidation rates; b) for seedlings: fresh and dry weight, shoot and root lengths, and total protein, soluble sugar, phenol and flavonoid contents. The dry weight was determined after drying the plant material in an oven at 40°C, until a constant weight.

Phytochemical Analysis

The hydroethanolic extracts were prepared with 250 mg of dry shoots and roots extracted with 10 mL of 70% ethanol (70% EtOH) in an ultrasound bath at 35°C for 30 min. This procedure was repeated 3 times. The extracts were dried at 45°C until the solvent was completely evaporated. The dry crude extract was then resuspended in 10 mL of 70% EtOH and used for the following determinations (Brazil, 2010).

Phenols were quantified with 100 µL of hydroethanolic extract according to Pastrana-Bonilla et al. (2003). The total phenol level was calculated using a calibration curve with 100 µg mL^{-1} gallic acid solution as the standard. Determinations were performed in triplicate, and the results were given in microgram equivalents of gallic acid per milligram of dry extract (µg GAEq mg^{-1} DE).

The total flavonoid assay was performed according to Woisky and Salatino (1998), and the flavonoid content was calculated using a calibration curve with 100 µg mL^{-1} quercetin in a methanol solution of 2% aluminum chloride as a standard. Determinations were performed in triplicate, and the results were given in microgram equivalents of quercetin per milligram of dry extract (µg QEq mg^{-1} DE).

Biochemical Analyses

Fresh shoots and roots (0.5 g) were mixed in 4 mL of ultrapure water. The samples were centrifuged at 4800 g for 30 min (Santos et al., 2010). The supernatants were collected to determine the total soluble sugar (Yemm and Willis, 1954) and total protein (Bradford, 1976) contents using colorimetric methods. The results were expressed in micrograms per milliliter of extract (µg/mL).

Data Analysis

Data were subjected to analysis of variance by an F test at a $p \leq 0.05$ significance level using the Variance Analysis System of Balanced Data SISVAR 5.1 Software (Ferreira, 2011). Mean rates were further separated by Scott–Knott's test when the differences were significant ($p < 0.05$).

Results

Freezing, *In Vitro* Germination and Phenotypic Analysis

Data on in vitro germination, contamination, oxidation and germination speed index (GSI) of *L. leucocephala* seeds are presented in Table 1. Depending on the treatment, the seeds had a reduced germination percentage of up to 62.5%. The highest percentage of germination and the lowest rates of contamination and oxidation were observed in the control. However, higher germination speed indexes (GSI) were recorded for seeds frozen for 21 days and thawed for up to 30 seconds (LC21a), which represented 68% germination.

Table 1. *In vitro* germination, contamination, oxidation and germination speed index (GSI) of *Leucaena leucocephala* seeds subjected to freezing in liquid nitrogen

Treatment	LN (days)	Thawing (seconds)	Germination (%)	Contamination (%)	Oxidation (%)	GSI
Control	-	-	96 a	4 d	0	10.33 b
LC7 a	7	30	60 c	32 b	8 d	9.67 b
LC7 b	7	60	56 c	24 c	20 b	10.50 b
LC7 c	7	90	56 c	32 b	12 c	9.50 b
LC7 d	7	120	64 c	24 c	12 c	10.83 b
LC14 a	14	30	76 b	4d	20 b	10.58 b
LC14 b	14	60	52 d	32 b	12 c	7.97 c
LC14 c	14	90	36 e	8 d	16 c	4.57 d
LC14 d	14	120	36 e	36 a	28 a	1.42 e
LC21 a	21	30	68 b	24 c	8 d	13.25 a
LC21 b	21	60	72 b	16 c	12 c	10.27 b
LC21 c	21	90	68 b	28 b	4 d	6.83 c
LC21 d	21	120	72 b	24 c	4 d	6.31 c

Means in each column followed by the same letter are not significantly different at $p < 0.05$ by the Scott-Knott's test.

Phenotypic analyses revealed that seedlings from seeds stored in LN for 21 days, regardless of the thawing time, showed high fresh and dry weights and shoot and root lengths (Table 2).

Phytochemical and Biochemical Analyses

According to Table 3, the levels of total phenols ranged from 1044 ± 54.7 to 658 ± 16.9 μg GAEq mg^{-1} DE in the shoots and from 528 ± 70.8 to 188 ± 20.7

μg GAEq mg^{-1} DE in the roots. In the shoots, the highest total phenol contents were found in the control, LC14a, LC14b, LC14d, LC21c and LC21d ($p < 0.05$). In the roots, the highest total phenol contents were observed in the LC7c and LC21d treatments ($p < 0.05$).

Table 2. Phenotypic analysis of 21-day-old *L. leucocephala* seedlings

Treatments	Shoots			Roots		
	Fresh Weight (g)	Dry Weight (g)	Length (cm)	Fresh Weight (g)	Dry Weight (g)	Length (cm)
Control	0.29 ± 0.05 a	0.04 ± 0.01 a	7.2 ± 0.5 b	0.12 ± 0.03 b	0.02* a	13.8 ± 2.4 c
LC7 a	0.23 ± 0.04 a	0.01* b	7.3 ± 1.5 b	0.11± 0.03 b	0.01* b	15.5 ± 2.7 b
LC7 b	0.27 ± 0.07 a	0.03 ± 0.01 a	6.9 ± 1.1 b	0.13 ± 0.04 b	0.01* b	16.8 ± 2.8 b
LC7 c	0.28 ± 0.05 a	0.03 ± 0.01 a	7.5 ± 0.6 b	0.12 ± 0.02 b	0.02 ± 0.01 a	17.2 ± 1.5 b
LC7 d	0.23 ± 0.06 a	0.03 ± 0.01 b	6.5 ± 1.8 b	0.09 ± 0.06 c	0.01* b	14.1 ± 4.9 c
LC14 a	0.25 ± 0.05 a	0.04 ± 0.01 a	7.4 ± 0.9 b	0.12 ± 0.04 b	0.02 ± 0.01 a	17.4 ± 1.9 b
LC14 b	0.23 ± 0.05 a	0.03 ± 0.01 a	5.0 ± 1.0 c	0.07 ± 0.04 c	0.01* b	12.0 ± 4.2 c
LC14 c	0.16 ± 0.07 a	0.02 ± 0.01 b	4.2 ± 2.5 c	0.06 ± 0.05 c	0.01* b	10.7 ± 8.9 c
LC14 d	0.10 ± 0.10 a	0.02 ± 0.01 b	1.6 ± 2.5 d	0.03 ± 0.04 d	0.004* c	4.5 ± 1.5 d
LC21 a	0.28 ± 0.10 a	0.04* a	8.8 ± 0.7 a	0.15 ± 0.05 a	0.02 ± 0.01 a	20.0 ± 4.2 a
LC21 b	0.27 ± 0.07 a	0.07 ± 0.09 a	8.2 ± 1.7 a	0.15 ± 0.05 a	0.02 ± 0.01 a	20.1 ± 6.3 a
LC21 c	0.29 ± 0.06 a	0.04 ± 0.01 a	8.5 ± 0.9 a	0.16 ± 0.04 a	0.02 ± 0.01 a	22.0 ± 2.0 a
LC21 d	0.30 ± 0.10 a	0.05 ± 0.01 a	8.5 ± 0.6 a	0.17 ± 0.02 a	0.03 ± 0.01 a	20.8 ± 3.2 a

The results were presented as the mean ± standard deviation. Means in each column followed by the same letter are not significantly different at $p < 0.05$ by the Scott-Knott's test. *Standard deviation less than 0.01.

A higher total flavonoid content was verified in the shoots for all treatments ($p < 0.05$) and ranged from 227 ± 5.3 to 156 ± 2.3 μg QEq mg^{-1} DE. In roots, the levels of total flavonoids were lower, in the order of 69 ± 1.1 to 46 ± 1.6 μg QEq mg^{-1} DE (Table 3).

Regarding the primary metabolites, the shoots and roots of the LC14b treatment (cryostored in liquid nitrogen for 14 days and thawed for 60 seconds) showed the highest contents of soluble sugars (1563 ± 27.8 μg/mL and 5017 ± 162.7 μg/mL, respectively) and total proteins (1413 ± 335.8 μg/mL and 8324 ± 274.0 μg/mL, respectively). We did not obtain enough material for the analysis of primary metabolites in the LC14c and LC14d treatments.

Table 3. Phytochemical (total phenols and flavonoids) and biochemical (soluble sugars and proteins) analyzes in shoots and roots of 21-days-old *L. leucocephala* seedlings

Shoots	Treatments	Total Phenols (μg GAEq mg^{-1} DE)	Total Flavonoids (μg QEq mg^{-1} DE)	Soluble Sugars (μg/mL)	Proteins (μg/mL)
	Control	1,044 ± 54.7 a	185 ± 1.9 a	489 ± 11.2 f	960 ± 78.3 e
	LC7 a	658 ± 16.9 b	186 ± 2.6 a	439 ± 23.5 f	570 ± 58.2 f
	LC7 b	739 ± 16.7 b	207 ± 3.8 a	383 ± 11.8 g	321 ± 107.5 g
	LC7 c	731 ± 24.4 b	186 ± 2.9 a	217 ± 1.4 g	289 ± 29.3 g
	LC7 d	837 ± 33.4 b	195 ± 4.8 a	575 ± 14.1 f	1,001 ± 132.1 d
	LC14 a	928 ± 24.2 a	195 ± 3.9 a	536 ± 4.7 f	594 ± 7.5 f
	LC14 b	975 ± 11.5 a	206 ± 2.0 a	1,563 ± 27.8 c	1,413 ± 335.8 d
	LC14 c	775 ± 32.5 b	184 ± 2.4 a	***	***
	LC14 d	1,030 ± 5.5 a	156 ± 2.3 b	***	***
	LC21 a	847 ± 19.1 b	207 ± 0.8 a	208 ± 12.3 g	95 ± 29.8 h
	LC21 b	875 ± 46.2 b	213 ± 2.6 a	238 ± 17.9 g	194 ± 53.7 h
	LC21 c	1,009 ± 50.4 a	227 ± 5.3 a	241 ± 18.2 g	393 ± 60.2 g
	LC21 d	931 ± 61.4 a	194 ± 2.7 a	242 ± 21.7 g	133 ± 9.8 h
Roots	Control	321 ± 23.8 d	67 ± 0.7 b	506 ± 71.8 f	1,437 ± 102.3 d
	LC7 a	317 ± 7.7 d	69 ± 1.1 b	884 ± 70.2 e	3,355 ± 134.3 b
	LC7 b	334 ± 2.6 d	60 ± 1.6 b	2,703 ± 151.9 b	2,794 ± 78.4 c
	LC7 c	498 ± 33.0 c	58 ± 0.8 b	1,820 ± 30.9 c	2,761 ± 129.1 c
	LC7 d	188 ± 20.7 e	67 ± 2.5 b	1,610 ± 47.4 c	1,730,1 ± 61.4 d
	LC14 a	365 ± 54.3 d	63 ± 0.3 b	1,096 ± 89.0 d	1,559 ± 14.0 d
	LC14 b	374 ± 19.3 d	57 ± 2.8 b	5,017 ± 162.7 a	8,324 ± 274.0 a
	LC14 c	377 ± 26.3 d	61 ± 1.2 b	***	***
	LC14 d	388 ± 25.1 d	58 ± 1.4 b	***	***
	LC21 a	361 ± 33.7 d	55 ± 1.6 b	1,121 ± 98.3 d	1,303 ± 154.9 d
	LC21 b	381 ± 11.9 d	58 ± 0.7 b	1,520 ± 154.3 c	3,046 ± 147.0 b
	LC21 c	365 ± 38.5 d	60 ± 0.1 b	556 ± 0.1 f	1,344 ± 42.3 d
	LC21 d	528 ± 70.8 c	46 ± 1.6 b	788 ± 64.8 e	1,368 ± 36.6 d

The results were presented as the mean ± standard deviation. Means in each column followed by the same letter are not significantly different at $p < 0.05$ by the Scott-Knott's test. GAEq = gallic acid equivalent; QEq = quercetin equivalent; DE = dry extract. *** There was no enough material for analysis.

Discussion

The control showed a germination rate of 96% (Table 1). The excision method used in this work was satisfactory for breaking tegument dormancy compared to the germination rate (98%) reported by Martinelli-Seneme et al. (2006) for *Bauhinia variegata* in a dormancy-breaking study; both species belong to the Fabaceae family. Seeds exposed to different cryostorage treatments in LN showed a lower germination rate compared to the control, possibly due to the crystallization of water contained in the seeds after immersion in LN, suggesting the need for adjustments in the time and/or temperature of thawing (Kaviani, 2011).

Exposure to LN imposes a series of stressors on plant material that are likely to induce changes in regenerated plants (Engelmann, 2011). There are few reports of phenotypic variations occurring during in vitro culture of cryostored plant materials, although Adu-Gyamfi, Wetten and López (2016) reported this type of variability, which may be epigenetic in nature and linked to *in vitro* culture conditions, cryopreservation and changes in DNA methylation. However, the existence of phenotypic variability is not a rule, as shown by Cejas et al. (2013). The results presented in this study confirm, at the phenotypic level, the effectiveness of cryostorage of *L. leucocephala* seeds to preserve and regenerate seedlings from *in vitro* culture, although experiments with a larger number of plants are needed to track potential changes.

A close relationship was not observed between the time of exposure of seeds to LN and the fresh and dry weights and shoot and root lengths, indicating that *L. leucocephala* seedlings are less affected by cryostorage and thawing compared to the control. Seeds stored for 21 days, regardless of the thawing time, showed a high accumulation of fresh and dry biomass and greater shoot and root lengths, with emphasis on seeds submitted to the LC21a treatment, which also presented a higher GSI (Tables 1 and 2).

However, this behavior was not similar for phytochemical and biochemical analyses. In general, a higher level of secondary metabolites (total phenol and flavonoids) was observed in the shoots, while roots preferentially accumulated primary metabolites (soluble sugars and proteins) (Table 3). Stress alters physiological and metabolic processes in plants (*Zevallos* et al., 2013). There are reports indicating that thawing after the exposure of embryos and seeds to very low temperatures may result in the alteration of essential metabolic functions (Prudente et al., 2019). The levels of primary and secondary metabolites in plants are subject to these changes and can be strongly resistant to stress conditions and environmental adaptations (Rosa et al., 2009). Knowledge about the biochemical and regulatory changes that occur during the *in vitro* growth of plant tissues can provide important information related to the establishment process and, consequently, provide the optimization of conditions for its in vitro culture (Mesquita et al., 2002).

Conclusion

Short-term liquid nitrogen storage is a tool that can be used for the conservation of *L. leucocephala* seeds. Seeds cryostored for 21 days showed a reduction in the germination rate; however, they maintained a high germination speed index and phenotypic characteristics in the seedlings. Higher levels of total phenols and flavonoids were observed in the shoots, while roots preferentially accumulated soluble sugars and proteins. These results highlight the biotechnological potential of *L. leucocephala* as an alternative source for the production of bioactive compounds and encourage additional studies to understand the mechanisms involved in the results described here.

Acknowledgments

We thank Dr. Andreia Fonseca Silva for the botanical identification of this plant. The authors are grateful to the Universidade Federal de São João del-Rei (UFSJ), Conselho Nacional de Desenvolvimento Científico e Tecnológico (CNPq), and Fundação de Amparo à Pesquisa do Estado de Minas Gerais for financial support. This study was financed in part by the Coordenação de Aperfeiçoamento de Pessoal de Nível Superior – Brasil (CAPES) - Finance Code 001.

Disclaimer

None.

References

Adu-Gyamfi, R., Wetten, A. and López, C. M. R. 2016. "Effect of cryopreservation and post-cryopreservation somatic embryogenesis on the epigenetic fidelity of cocoa (*Theobroma cacao* L.)." *Plos One*, 11(7):e0158857. doi: 10.1371/journal.pone.0158857.

Alves, M. C. S., Filho, S. M., Andrade-Neto, M. and Teófilo, E. M. 2000. "Superação da dormência em sementes de *Bauhinia monadra* Britt. e *Bauhinia ungulata* L. – Caesalpinoideae" [Overcoming dormancy in *Bauhinia monadra* Britt seeds. and

Bauhinia ungulata L. – Caesalpinoideae]. *Revista Brasileira de Sementes*, 22(2):139-144.

Bradford, M. M. 1976. "A rapid and sensitive method for the quantification of microgram quantities of protein utilizing the principle of protein-dye binding." *Analytical Biochemistry*, 72:248-254.

Brazil. 2010. *Brazilian Pharmacopeia.* Brasília: Anvisa.

Burawat, J., Uabandit, N., Sripanidkulchai, B., Nualkaew, S. and Iamsaard, S. 2016. "Antioxidant capacity and acute testicular toxicity of *Leucaena leucocephala* aqueous shoot tips plus young leaves extracts." *Int. J. Morphol.*, 34(2):514-521.

Cejas, I., Méndez, R., Villalobos, A., Palau, F., Aragón, C., Engelmann, F., Carputo, D., Aversano, R., Martínez, M. E. and Lorenzo, J. C. 2013. "Phenotypic and molecular characterization of *Phaseolus vulgaris* plants from non-cryopreserved and cryopreserved seeds." *American Journal of Plant Sciences*, 4:844-849. doi: 10.4236/ajps.2013.44103.

Chen, X. L., Li, J. H., Xin, X., Zhang, Z. E., Xin, P. P. and Lu, X. X. 2011. "Cryopreservation of *in vitro*-grown apical meristems of *Lilium* by droplet-vitrification." *South African Journal of Botany*, 77:397-403. doi: 10.1016/j.sajb.2010.10.005.

Engelmann, F. 2011. "Use of biotechnologies for the conservation of plant biodiversity." *In vitro & Developmental Biology*, 47:5-16. doi: 10.1007/s11627-010-9327-2.

Ferreira, D. F. 2011. "Sisvar: a computer statistical analysis system." *Ciência e Agrotecnologia*, 35(6):1039-1042.

Kaviani, B. 2011. "Conservation of plant genetic resources by cryopreservation." *Australian Journal of Crop Science*, 5(6):778-800.

Li, D. Z. and Pritchard, H. W. 2009. "The science and economics of ex situ plant conservation." *Trends Plant Sci*, 14:614–621. doi: 10.1016/j.tplants.2009.09.005.

Maguire, J. D. 1962. "Speed of germination aid in selection and evaluation for seedling emergence and vigor." *Crop Science*, 2(1):176-177.

Martinelli-Seneme, A., Possamai, E., Schuta, L. R. and Vanzolini, S. 2006. "Germinação e sanidade de sementes de *Bauhinia variegata*." [Germination and sanity of *Bauhinia variegata* seeds]. *Revista Árvore*, 30(5):719-724. doi: 10.1590/s0100-67622006000500005.

Mesquita, A. C., Paiva, R., Santiago, E. J. A., Paiva, P. D. O., Paiva, L. V. and Gomes, G. A. C. "Análises bioquímicas de calos obtidos de segmentos foliares de lichieira (*Litchi chinensis* Sonn.)." [Biochemical analyzes of calluses obtained from lychee leaf segments (*Litchi chinensis* Sonn]. *Magistra*, 14(1):1-6.

Mittal, N., Mattu, P. and Kaur, G. 2016. "Extraction and derivatization of *Leucaena leucocephala* (Lam.) galactomannan: Optimization and characterization." *International Journal of Biological Macromolecules*, 92:831-841. doi: 10.1016/j.ijbiomac.2016.07.046.

Murashige, T. and Skoog, F. 1962. "A revised medium for rapid growth and bioassays with tobacco tissue cultures." *Physiologia Plantarum, Copenhagen*, 15(3): 473-497.

Nascimento, P. K. V., Franco, E. T. H and Frassetto, E. G. 2007. "Desinfestação e germinação *in vitro* de sementes de *Parapiptadenia rigida* Bentham (Brenam)."

[Disinfestation and *in vitro* germination of *Parapiptadenia rigida* Bentham (Brenam) seeds]. *Revista Brasileira de Biociências*, 5(2): 141-143.

Pastrana-Bonilla, E., Akoh, C. C., Sellappan, S. and Gerard, K. 2003. "Phenolic Content and Antioxidant Capacity of Muscadine Grapes." *Journal Agricultural and Food Chemistry* 51:5497-503. doi: 10.1021/jf030113c.

Pêgo, R. G., Paiva, P. D. O. and Paiva, R. 2013. "Micropropagation of *Syngonanthus elegantulus." Ciência e Agrotecnologia*, 37(1): 32-39.

Pires, L. M., Bragantini, C. and Costa, J. L. S. 2004. "Armazenamento de sementes de feijão revestidas com polímeros e tratadas com fungicidas." [Storage of polymer-coated and fungicide-treated bean seeds]. *Pesquisa Agropecuária Brasileira*, 39(7): 709-715.

Prudente, D. O., Paiva, R., Domiciano, D., de Souza, L. B., Carpentier S., Swennen, R., Silva, L. C., Nery, F. C., Máximo, W. P. F. and Panis, B. 2019. "The cryoprotectant PVS2 plays a crucial role in germinating *Passiflora ligularis* embryos after cryopreservation by influencing the mobilization of lipids and the antioxidant metabolism." *Journal of Plant Physiology*, 239:71–82. doi: 10.1016/j.jplph.2019.05.014.

Rosa, M., Prado, C., Podazza, G., Interdonato, R., González, J. A., Hilal, M. and Prado, F. E. 2009. "Soluble sugars - Metabolism, sensing and abiotic stress." *Plant Signaling & Behavior*, 4(5):388-393. doi: 10.4161/psb.4.5.8294.

Santos, D. N., Nunes, C. F., Pasqual, M., Valente, T. C. T., Oliveira, A. C. L. and Silveira, N. M. 2010. "Análises bioquímicas em calos de pinhão-manso." [Biochemical analyzes in *Jatropha* callus]. *Ciência Rural*, 40(11):2268-2273.

Soares, A. M. S., Araujo, S. A., Lopes, S. G. and Junior, L. M. C. 2015. "Anthelmintic activity of *Leucaena leucocephala* protein extracts on *Haemonchus contortus*." *Revista Brasileira de Parasitologia Veterinária*, 24(4):396-401. doi: 10.1590/s1984-29612015072.

Teles, M. M., Alves, A. A., Oliveira, J. C. G. and Bezerra, A. M. E. 2000. "Métodos para Quebra da Dormência em Sementes de Leucena (*Leucaena leucocephala* (Lam.) de Wit"). [Methods for breaking dormancy in seeds of *Leucaena* (*Leucaena leucocephala* (Lam.) de Wit]. *Revista Brasileira de Zootecnia*, 29(2):387-391.

Verma, V. K., Rani, K. V., Kumar, S. R. and Prakash, O. 2018. "*Leucaena leucocephala* pod seed protein as an alternate to animal protein in fish feed and evaluation of its role to fight against infection caused by *Vibrio harveyi* and *Pseudomonas aeruginosa*." *Fish & Shellfish Immunology*, 76:324-332. doi: 10.1016/j.fsi.2018.03.011.

Woisky, R. G. and Salatino, A. 1998. "Analysis of Propolis: Some Parameters and Procedures for Chemical Quality Control." *Journal of Apicultural Research* 37:99-105. doi: 10.1080/00218839.1998.11100961.

Yemm, E. W. and Willis, A. J. 1954. "The estimation of carbohydrates in plant extracts by anthrone." *The Biochemical Journal*, 57:508-514.

Zevallos, B., Cejas, I., Rodríguez, R. C., Yabor, L., Aragón, C., González, J., Engelmann, F., Martínez, M. E. and Lorenzo, J. C. 2013. "Biochemical characterization of ecuadorian wild *Solanum lycopersicum* Mill. plants produced from non-cryopreserved and cryopreserved seeds." *Cryo Letters*, 34(4):413-421.

Chapter 11

Sphingosine-1-phosphate and Its Receptors Modulate Changes of Macrophages during Stimulation

Zihan Deng[1,*]
Tingwen Xiang[2,†]
and Yueqi Chen[1,‡]
[1]Department of Orthopedics, Southwest Hospital, Third Military Medical University (Army Medical University), Chongqing, People's Republic of China
[2]College of Basic Medical Science, Third Military Medical University (Army Medical University), Chongqing, People's Republic of China

Abstract

Sphingosine-1-phosphate (S1P) acts as a bioactive sphingolipid, synthesized by the actions of sphingosine kinases (SphKs). S1P is ubiquitously expressed in the human body by binding to G-protein-coupled receptors S1P receptors (S1PRs) and plays a dramatic role in the immune system. Interestingly, in the event of aberrant pathological changes (such as inflammation, injury, etc.) occurring in tissues, S1P levels are dramatically increased, which can affect the recruitment of immune cells including macrophages. With the increasing attention of researchers, several researches have been made to figure out the promising mechanisms by modulating macrophages function via S1P and its receptors. In the last few years, accumulating evidence suggests that S1P-S1PRs signaling pathway in macrophages plays a central role in the

* The author contributed equally to this work.
† The author contributed equally to this work.
‡ Corresponding Author's Email: chenyueqi1012@sina.com.

In: Advances in Biology. Volume 4
Editor: Charles D. Grant
ISBN: 979-8-88697-999-2

pathogenesis of inflammatory diseases such as atherosclerosis, asthma, rheumatoid arthritis (RA), and multiple sclerosis (MS). In this chapter, we review the biological properties of S1P and summarize the current knowledge of the function about S1P in macrophage biology. The elaboration of these alterations and implications will facilitate the development of potential strategies for treating different diseases based on the synergy between S1P-S1PRs axis and macrophages.

Keywords: sphingosine-1-phosphate (S1P), S1P receptors (S1PRs), macrophages, inflammation

Introduction

Macrophages are ubiquitous heterogeneous population of immune cells, which have been implicated in a number of physiological and pathological processes, due to their high plasticity in functional responses [1]. As innate immune cells, macrophages are present in almost all tissues, and their indispensable functions runs through critical stages of both innate and adaptive immunity [2]. Macrophages maintain normal tissue homeostasis and protect against immune-mediated pathological damage by phagocytosis of invading microorganisms, increased inflammatory responses, as well as recruitment of additional immune cells [3, 4]. In normal tissues, the phenotypic and functional properties of macrophages can be changed in response to various microenvironment signals, including diverse cytokines and other mediators. Therefore, macrophages have long been considered as a series of unstable subgroups of cells [3]. Traditionally, there are two distinct phenotypes of macrophages, namely classically activated macrophages (M1) and alternatively activated macrophages (M2) [5].

Recently, increasing attention has been paid to macrophages for their pivotal roles in immune response, tissue repair, and maintaining normal tissue homeostasis. And in order to successfully perform above functions, a number of molecules have been mastered by macrophages to recognize and indicate endangered or disturbed microenvironment homeostasis, one of which is sphingosine-1-phosphate (S1P), that is gradually produced during inflammation and tissue damage. S1P is a bioactive sphingolipid metabolite which participates in a number of important physiological processes such as angiogenesis and immune responses [6, 7]. Meanwhile, S1P also play as a well-known secondary messenger in the inflammatory process, coordinating signal transduction [8, 9]. Produced by the phosphorylation of sphingosine via two distinct sphingosine kinases (SphKs), SphK1 and SphK2, S1P exports out of cells and transmits information, via different transporters such as ATP-binding cassette transporters and spinster homolog 2 (Spns2). Numerous actions of S1P in diverse physiological and pathological processes are mediated by binding to specific G protein-coupled cell surface receptors. And there are currently five known subtypes of S1P receptors (S1PRs), consisting of S1PR1 to S1PR5 [10]. To date, S1P production, S1PRs expression, and signaling mechanisms mediated by both have been identified to contribute to the regulation of macrophages, but the specific analysis is still challenging. In this chapter, we have summarized the biological characteristics and functions of S1P, and describe recent advances in how S1P and its diverse receptors regulate macrophages function as well as further affect immune responses, which is expected to provide promising therapeutic strategies for treating different diseases.

Macrophage Diversity and Function

Present in almost all tissues as immune cells, macrophages originate from myeloid progenitor cells in the bone marrow, and differentiate into monocytes mediated by cytokines such as granulocyte macrophage colony-stimulating factor. Subsequently, monocytes enter the bloodstream, undergo chemotactic reactions under appropriate stimuli, and migrate to different tissues. Then, under the influence of growth factors such as macrophage colony-stimulating factor, monocytes differentiate into tissue macrophages, and their functions are dependent on the respective tissues microenvironment in which they reside [11, 12].

As a part of the innate immune system, macrophages play dramatic role in maintaining normal tissue homeostasis and protecting against immune-mediated pathological damage. Acting as prodigious phagocytic cells, they can effectively eliminate cells which have experienced different patterns of cell death. Moreover, macrophages can migrate along the gradient of chemokines, and prevent the dynamic balance of tissue microenvironment from being disrupted via sensing and intervening inflammatory events. Notably, macrophages play an integral role in connecting innate and adaptive immune responses through secreting diverse cytokines and presenting endogenous or exogenous antigens [11].

As is well known, macrophages represent a range of activated phenotypes that exhibit remarkable plasticity. In general, macrophages can be polarized into two distinct subtypes, and researchers usually refer to classically activated macrophages as M1 macrophages, and alternatively activated macrophages as M2 macrophages [5]. Generally considered to be stimulated with interferon-γ (IFN-γ) and microbial products like lipopolysaccharide (LPS), M1 macrophages are characterized to display an immune protective process during type I inflammation (defense against microbial infection), whereas M2 macrophages, which are induced by interleukin-4 (IL-4), IL-10, IL-13, glucocorticoids, and immunoglobulin complexes, mimic conditions of type II inflammation (helminth infection) [13]. M1 macrophages are considered to be potent anti-microbial and pro-inflammatory cells, and their transcriptional characteristics are demonstrated by the activation of nuclear factor kappa-light-chain-enhancer of activated B cells (NF-κB), signal transducer and activator of transcription 1 (STAT1), as well as interferon regulatory factor 5 (IRF5) [14, 15]. M1 macrophages possess high antigen-presenting capacity, and can generate pro-inflammatory cytokines such as tumor necrosis factor-α (TNF-α), IL-1β, IL-6, and IL-12, and also promote the generation of nitric oxide (NO) and reactive oxygen species (ROS), as a result facilitating type I immune response to mediate host defence against infections of bacteria and viruses. In contrast, M2 macrophages are activated by transcription factors STAT6 and IRF4, expressing specific chemokines including CCL17 and CCL18, as well as phagocytic receptors like mannose-receptor CD206, along with multiple mediators that regulate the extracellular matrix [16]. All of these are induced to modulate tissue remodeling, facilitate adaptive Th2 immunity, and combat extracellular parasites [15, 17-19].

Despite it is generally accepted that the microenvironment in which macrophages are located triggers transcriptional responses, thus shaping their specific phenotypes and function, the underlying mechanisms that guide the polarization of macrophages have not yet been fully clarified [20]. Proper management of macrophage-mediated inflammatory response and dynamic maintenance of M1 and M2 macrophages balance are essential for sustaining the homeostasis of the host. Accumulating evidence suggests that activated macrophages participate in a number of disease processes, such as autoimmune diseases, allergic disorders, cancer, and infections. Therefore, recognizing macrophage activation states and modulating the macrophage polarization from M1 to M2 or vice versa may provide new diagnostic approaches or therapeutic strategies for a variety of diseases.

S1P and Its Receptors

S1P, a bioactive sphingolipid, is generated in cytosol and nucleus of diverse cells via the phosphorylation of sphingosine by an enzyme family consisting of two isoenzymes, namely SphK1 and SphK2. As a multifunctional signaling lipid molecule, S1P not only plays as the second messenger within cells, which directly acts on intracellular targets, participating in signal transduction processes, but is also able to be exported from cells via ATP-binding cassette transporters or Spns2, combining and activating S1PRs on the cell membrane surface. Once S1P is transferred extracellular, it can bind to one of the five distinct G protein-coupled receptors (S1PR1-S1PR5), and transmit signals through autocrine or paracrine manners [20, 21]. Subsequently, S1P complex is capable of triggering a range of downstream signaling pathways, mediating a variety of cellular functions, such as cell proliferation, migration, survival, apoptosis as well as communication, and plays an important role in immune modulation, hematopoietic regulation, inflammation management, and other biological effects [22, 23].

As members of a family of seven helix transmembrane G protein-coupled receptors, S1PRs can be attached to G_i, G_q, and/or $G_{12/13}$. The diversity of S1P biological effects mainly depends on the distinct subtypes of coupled protein receptors and the specific expression of G protein receptors in cells [24]. At present, five subtypes of S1PRs have been found, namely S1PR1 to S1PR5, and S1P is capable of binding to these five receptors at low concentration (nmol/L level). Researches have shown that S1PRs are distributed differently in diverse tissues and cell types, thus mediating multiple biological effects and involved in the regulation of essential physiological or pathological processes, which are centrally expressed in the immune and vascular systems. Among these receptors, S1PR1, S1PR2 and S1PR3 are widely expressed in different tissues in the human body. During embryonic development stage, S1PR1 is highly expressed in the vascular system, which is particularly significant for the normal development of the vasculature. And in adults, S1PR1 is expressed in a variety of tissues and organs, such as brain, lung, heart, liver, and so on, which is an important receptor on the surface of lymphocytes, mediating the homing of lymphocytes. In addition, S1P-S1PR1 signaling pathway plays an indispensable role in chemotaxis, activation and differentiation of immune cells, and thus has long been regarded as a key regulatory factor of immune response [10, 25-28]. Moreover, S1PR2 is involved in mast cell degranulation and histamine secretion, as well as inhibits the movement of specific cells by the suppression of chemotactic receptors [29, 30]. Intriguingly, S1PR3 regulates S1P-induced vasoconstriction, and seems to be crucial for inflammatory and fibrotic responses [31-33]. Furthermore, recent evidences have indicated that S1P4 is mainly expressed in lymphatic and hematopoietic tissues, while S1P5 may play a role in central nervous system related pathways. Due to the expression of S1PR4 and S1PR5 is relatively limited and only detected in specific tissues, there are currently few studies on SIPR4 and S1PR5.

It is worth noting that macrophages can express all five S1PRs, and the activation of different receptors seems to be related to their different functional roles [16, 34]. S1P binds to distinct receptors on macrophages, thus causing directly related signaling pathways to be affected, such as small GTPases, phospholipases, PI3K or adenylyl cyclase, which initiates a number of different signals. As a typical sensitive signal caused by the disruption of dynamic equilibrium of tissues, S1P cannot activate specific receptors due to its relatively low levels under normal conditions, while in the process of inflammation, S1P levels change and are successfully sensed by macrophages [35]. As scavengers of the immune system, macrophages have long been considered to be active in different tissue microenvironments to receive diverse signal stimuli, which enables them to adjust their activation states and polarization patterns under physiological or pathological conditions. However, the production of S1P, the expression of S1PRs, as well as the signal transduction of S1P-S1PRs axis may lead to the complexity and diversity of the functional properties of macrophages. Therefore, it is of great significance to explore whether modulating the motility of macrophages via S1P and its receptors exhibits potential therapeutic effects.

Synergistic Effects of S1P and Its Receptors on Macrophages

Accumulating evidence suggests that the effects of S1P on macrophage phenotype transition have gradually become so pivotal. S1P is able to regulate the macrophages polarization and repolarization in response of the changes of local microenvironment, as well as the activated S1PRs on the cells. Thereby, on the one hand, S1P can facilitate the generation of M2 macrophages, and on the other hand, it is also able to induce the production of M1 macrophages associated markers. Previous studies have shown that S1P not only modulates the polarization of macrophages, but also can affect tissue homeostasis mediated by intervening in macrophage differentiation, migration, and survival, thereby exerting a dramatic effect in immune surveillance. In addition, S1P is able to act as an anti-apoptotic signal in macrophages too. Based on current researches, S1P can either activate phosphoinositide-3-kinase (PI3K), extracellular-signal regulated kinase (ERK) and Ca^{2+} signaling pathways, or alter the cellular balance of ceramide to sphingosine or S1P, thus inducing the expression of anti-apoptotic proteins including B-cell lymphoma 2 (Bcl2) and B-cell lymphoma extra-large (Bcl-XL), ultimately preventing caspase-induced macrophage apoptosis [16].

Besides, in order to keep the dynamic balance of tissues, macrophages not only have to modulate their numbers, but also need to appear in the proper place at the appropriate time. Hence, S1P is capable of playing as a lipid attraction signal, which correctly directs macrophages to the major regions of inflammatory erosion and tissue repair [36]. Notably, a large number of comprehensive studies have demonstrated that S1P plays an important role in the process of S1P-mediated macrophage migration, which is rigorously dependent on their S1PRs profile.

Regarding the S1PR1 signaling pathway, many studies have suggested that it appears to be pro-migratory, as well as has been indicated to play a key role in the migration of macrophages following the onset of inflammation. For example, *Weichand, B* et al. have found that in the mouse models of peritoneal inflammation, the specific deletion of S1PR1 on the surface of macrophages results in markedly reduced migration of macrophages from the site of inflammation [37]. Instead, S1PR4 signaling may restrict the pro-migratory effect of S1PR1. It has been shown that the S1PR1/S1PR4 ratio is higher in pro inflammatory M1 macrophages than that in anti-inflammatory M2 macrophages in an inflammatory context, thereby prompting that the S1PR1/S1PR4 ratio exerts a crucial effect in the migration of M1 macrophages from sites of inflammation [16]. Nevertheless, further relevant researches such as respective inhibition of S1PR1 and S1PR4 in macrophages are needed to elucidate the functional effect of S1PR4 in the reverse modulation of S1PR1 during inflammation. Additionally, based on *Michaud J* et al. studies of S1PR2 knockout mice in an acute inflammatory peritonitis model, it has demonstrated that S1PR2 signaling inhibits macrophage transport via the stimulation of producing cAMP, thus reducing the phosphorylation of Akt, and then further indicating that S1PR1 and S1PR2 have synergistic effects in the migration of macrophages [29]. Furthermore, it is worth noting that S1PR3 has been shown to play a central role in the activation of pro-inflammatory macrophages through the utilization of murine microglia in vitro and a model of brain ischemia. The depletion of S1PR3 induced by the intervention of S1PR3 specific antagonists and siRNA can reduce the expression of LPS-induced pro-inflammatory factors such as TNF-α, IL-6, and IL-1β [38]. Interestingly, in macrophages of primary peritonitis, S1PR3 antagonists can also decrease the expression of LPS-induced inflammatory genes, thus reducing the release of pro-inflammatory mediators including TNF-α, IL-6, IL-1β, iNOS, and COX-2 [39]. As a result, it's reasonable to speculate that S1PR3 seems to act as an inflammatory receptor in macrophages. In addition, it is well known that S1PR4 and S1PR5 are less expressed in most macrophages, so their effects on macrophage polarization and migration remain less well defined.

Of note, macrophages are cells with high plasticity in the immune system, and during their migration, multiple roles regarding the function of specific S1P receptors may depend on different S1PRs expression profiles and the microenvironment in which macrophages reside. To sum up, in order to maintain the dynamic balance of tissues, macrophages need to be activated in situations of stress and rapidly migrate to specific tissue sites. The above studies have indicated that the S1P-S1PRs signal axis may play an indispensable role in these effects.

Conclusion

Macrophages, as important immune cells, can accurately perceive and actively deal with the disturbance of tissue homeostasis caused by changes in the microenvironment through rapidly initiating and resolving the inflammatory processes, which play a dramatic role in maintaining the dynamic balance of tissue microenvironment. S1P is a sensitive multifunctional signaling molecule that is found at low levels in tissues except blood and lymph. Thus, slight changes in S1P levels signify alterations in the tissue microenvironment, and are promptly sensed by macrophages and their progenitors, which induce alterations in macrophage migration, polarization, and effector function, participating in many regulatory processes including immune surveillance. The source of S1P and the type of S1PRs are key determinants by which S1P affects and regulates macrophages.

In summary, the S1P system appears to be a strong modulator of macrophage function, and regulates diverse biological functions via the S1P-S1PRs axis, which is critical in physiological and pathological processes of cardiovascular, immune, or central nervous system. Defects in S1PRs dependent signaling may lead to the development of many diseases, such as atherosclerosis, autoimmune diseases, or cancer. Considering the differential expression profiles of S1PRs on macrophages under certain pathological conditions, targeted therapies (agonists and antagonists) against different S1PRs, which modulate the pathophysiological functions of macrophages, will serve as a potential therapeutic strategy to alleviate the progression of different diseases in the future.

References

[1] Moin, A. S. M., Sathyapalan, T., Diboun, I., Atkin, S. L., Butler, A. E. (2021) Identification of macrophage activation-related biomarkers in obese type 2 diabetes that may be indicative of enhanced respiratory risk in COVID-19. *Scientific reports* 11:6428.

[2] Galli, S. J., Borregaard, N., Wynn, T. A. (2011) Phenotypic and functional plasticity of cells of innate immunity: macrophages, mast cells and neutrophils. *Nat Immunol* 12:1035-1044.

[3] Murray, P. J., Wynn, T. A. (2011) Protective and pathogenic functions of macrophage subsets. *Nature reviews. Immunology* 11:723-737.

[4] Hu, Y., Huang, J., Chen, C., Wang, Y., Hao, Z., Chen, T., Wang, J., Li, J. (2022) Strategies of Macrophages to Maintain Bone Homeostasis and Promote Bone Repair: A Narrative Review. *Journal of functional biomaterials* 14.

[5] Mosser, D. M., Edwards, J. P. (2008) Exploring the full spectrum of macrophage activation. *Nature reviews. Immunology* 8:958-969.

[6] Takabe, K., Paugh, S. W., Milstien, S., Spiegel, S. (2008) "Inside-out" signaling of sphingosine-1-phosphate: therapeutic targets. *Pharmacological reviews* 60:181-195.

[7] Takabe, K., Spiegel, S. (2014) Export of sphingosine-1-phosphate and cancer progression. *Journal of lipid research* 55:1839-1846.

[8] Spiegel, S., Milstien, S. (2011) The outs and the ins of sphingosine-1-phosphate in immunity. *Nature reviews. Immunology* 11:403-415.

[9] Huang, W. C., Nagahashi, M., Terracina, K. P., Takabe, K. (2013) Emerging Role of Sphingosine-1-phosphate in Inflammation, Cancer, and Lymphangiogenesis. *Biomolecules* 3:408-434.

[10] Kihara, Y., Maceyka, M., Spiegel, S., Chun, J. (2014) Lysophospholipid receptor nomenclature review: IUPHAR Review 8. *British journal of pharmacology* 171:3575-3594.

[11] Gordon, S. (2003) Alternative activation of macrophages. *Nature reviews. Immunology* 3:23-35.

[12] Stout, R. D., Jiang, C., Matta, B., Tietzel, I., Watkins, S. K., Suttles, J. (2005) Macrophages sequentially change their functional phenotype in response to changes in microenvironmental influences. *Journal of immunology (Baltimore, Md. 1950)* 175:342-349.

[13] Zhu, L., Yang, T., Li, L., Sun, L., Hou, Y., Hu, X., Zhang, L., Tian, H., Zhao, Q., Peng, J., Zhang, H., Wang, R., Yang, Z., Zhang, L., Zhao, Y. (2014) TSC1 controls

macrophage polarization to prevent inflammatory disease. *Nature communications* 5:4696.

[14] Krausgruber, T., Blazek, K., Smallie, T., Alzabin, S., Lockstone, H., Sahgal, N., Hussell, T., Feldmann, M., Udalova, I. A. (2011) IRF5 promotes inflammatory macrophage polarization and TH1-TH17 responses. *Nat Immunol* 12:231-238.

[15] Hu, S., Liu, H., Ha, Y., Luo, X., Motamedi, M., Gupta, M. P., Ma, J. X., Tilton, R. G., Zhang, W. (2015) Posttranslational modification of Sirt6 activity by peroxynitrite. *Free radical biology & medicine* 79:176-185.

[16] Weigert, A., Olesch, C., Brüne, B. (2019) Sphingosine-1-Phosphate and Macrophage Biology-How the Sphinx Tames the Big Eater. *Frontiers in immunology* 10:1706.

[17] Murray, P. J., Allen, J. E., Biswas, S. K., Fisher, E. A., Gilroy, D. W., Goerdt, S., Gordon, S., Hamilton, J. A., Ivashkiv, L. B., Lawrence, T., Locati, M., Mantovani, A., Martinez, F. O., Mege, J. L., Mosser, D. M., Natoli, G., Saeij, J. P., Schultze, J. L., Shirey, K. A., Sica, A., Suttles, J., Udalova, I., van Ginderachter, J. A., Vogel, S. N., Wynn, T. A. (2014) Macrophage activation and polarization: nomenclature and experimental guidelines. *Immunity* 41:14-20.

[18] Brüne, B., Dehne, N., Grossmann, N., Jung, M., Namgaladze, D., Schmid, T., von Knethen, A., Weigert, A. (2013) Redox control of inflammation in macrophages. *Antioxidants & redox signaling* 19:595-637.

[19] Martinez, F. O., Helming, L., Milde, R., Varin, A., Melgert, B. N., Draijer, C., Thomas, B., Fabbri, M., Crawshaw, A., Ho, L. P., Ten Hacken, N. H., Cobos Jiménez, V., Kootstra, N. A., Hamann, J., Greaves, D. R., Locati, M., Mantovani, A., Gordon, S. (2013) Genetic programs expressed in resting and IL-4 alternatively activated mouse and human macrophages: similarities and differences. *Blood* 121:e57-69.

[20] Lawrence, T., Natoli, G. (2011) Transcriptional regulation of macrophage polarization: enabling diversity with identity. *Nature reviews. Immunology* 11:750-761.

[21] Rivera, J., Proia, R. L., Olivera, A. (2008) The alliance of sphingosine-1-phosphate and its receptors in immunity. *Nature reviews. Immunology* 8:753-763.

[22] Rosen, H., Goetzl, E. J. (2005) Sphingosine 1-phosphate and its receptors: an autocrine and paracrine network. *Nature reviews. Immunology* 5:560-570.

[23] Xiao, L., Zhou, Y., Friis, T., Beagley, K., Xiao, Y. (2019) S1P-S1PR1 Signaling: the "Sphinx" in Osteoimmunology. *Frontiers in immunology* 10:1409.

[24] Mendelson, K., Evans, T., Hla, T. (2014) Sphingosine 1-phosphate signalling. *Development (Cambridge, England)* 141:5-9.

[25] Yanagida, K., Hla, T. (2017) Vascular and Immunobiology of the Circulatory Sphingosine 1-Phosphate Gradient. *Annual review of physiology* 79:67-91.

[26] Obinata, H., Hla, T. (2019) Sphingosine 1-phosphate and inflammation. *International immunology* 31:617-625.

[27] Mendoza, A., Fang, V., Chen, C., Serasinghe, M., Verma, A., Muller, J., Chaluvadi, V. S., Dustin, M. L., Hla, T., Elemento, O., Chipuk, J. E., Schwab, S. R. (2017) Lymphatic endothelial S1P promotes mitochondrial function and survival in naive T cells. *Nature* 546:158-161.

[28] Aoki, M., Aoki, H., Ramanathan, R., Hait, N. C., Takabe, K. (2016) Sphingosine-1-Phosphate Signaling in Immune Cells and Inflammation: Roles and Therapeutic Potential. *Mediators of inflammation* 2016:8606878.

[29] Michaud, J., Im, D. S., Hla, T. (2010) Inhibitory role of sphingosine 1-phosphate receptor 2 in macrophage recruitment during inflammation. *Journal of immunology (Baltimore, Md. : 1950)* 184:1475-1483.

[30] Laidlaw, B. J., Gray, E. E., Zhang, Y., Ramírez-Valle, F., Cyster, J. G. (2019) Sphingosine-1-phosphate receptor 2 restrains egress of γδ T cells from the skin. *The Journal of experimental medicine* 216:1487-1496.

[31] Cantalupo, A., Gargiulo, A., Dautaj, E., Liu, C., Zhang, Y., Hla, T., Di Lorenzo, A. (2017) S1PR1 (Sphingosine-1-Phosphate Receptor 1) Signaling Regulates Blood Flow and Pressure. *Hypertension (Dallas, Tex. : 1979)* 70:426-434.

[32] Murakami, K., Kohno, M., Kadoya, M., Nagahara, H., Fujii, W., Seno, T., Yamamoto, A., Oda, R., Fujiwara, H., Kubo, T., Morita, S., Nakada, H., Hla, T., Kawahito, Y. (2014) Knock out of S1P3 receptor signaling attenuates inflammation and fibrosis in bleomycin-induced lung injury mice model. *PloS One* 9:e106792.

[33] Takuwa, N., Ohkura, S., Takashima, S., Ohtani, K., Okamoto, Y., Tanaka, T., Hirano, K., Usui, S., Wang, F., Du, W., Yoshioka, K., Banno, Y., Sasaki, M., Ichi, I., Okamura, M., Sugimoto, N., Mizugishi, K., Nakanuma, Y., Ishii, I., Takamura, M., Kaneko, S., Kojo, S., Satouchi, K., Mitumori, K., Chun, J., Takuwa, Y. (2010) S1P3-mediated cardiac fibrosis in sphingosine kinase 1 transgenic mice involves reactive oxygen species. *Cardiovascular research* 85:484-493.

[34] Bryan, A. M., Del Poeta, M. (2018) Sphingosine-1-phosphate receptors and innate immunity. *Cellular microbiology* 20:e12836.

[35] Hla, T., Venkataraman, K., Michaud, J. (2008) The vascular S1P gradient-cellular sources and biological significance. *Biochimica et biophysica acta* 1781:477-482.

[36] Gude, D. R., Alvarez, S. E., Paugh, S. W., Mitra, P., Yu, J., Griffiths, R., Barbour, S. E., Milstien, S., Spiegel, S. (2008) Apoptosis induces expression of sphingosine kinase 1 to release sphingosine-1-phosphate as a "come-and get-me" signal. *FASEB journal : official publication of the Federation of American Societies for Experimental Biology* 22:2629-2638.

[37] Weichand, B., Weis, N., Weigert, A., Grossmann, N., Levkau, B., Brüne, B. (2013) Apoptotic cells enhance sphingosine-1-phosphate receptor 1 dependent macrophage migration. *European journal of immunology* 43:3306-3313.

[38] Gaire, B. P., Song, M. R., Choi, J. W. (2018) Sphingosine 1-phosphate receptor subtype 3 (S1P(3)) contributes to brain injury after transient focal cerebral ischemia via modulating microglial activation and their M1 polarization. *Journal of neuroinflammation* 15:284.

[39] Heo, J. Y., Im, D. S. (2019) Pro-Inflammatory Role of S1P(3) in Macrophages. *Biomolecules & therapeutics* 27:373-380.

Chapter 12

The Structure of the Neutrino Halo (Bio Field) of a Chicken Egg

Kristina Zubow[1,*]
Anatolij Zubow[2]
and Viktor Zubow[1]
[1]R&D, Zubow Consulting (NPO), Peenehagen, Germany
[2]Department of Telecommunication Networks Group, TU, Berlin, Germany

Abstract

Gravitational mass spectroscopy (GMS) was used to study the structure of the neutrino halo (bio field) around an unfertilized egg of a domestic hen (*Gallus gallus domesticus*). The presence of a halo was established, which extends to a distance of 50 cm from the surface of the shell. The halo is not homogeneous. It is a superposition of standing shock waves of neutrinos from proton oscillations in atomic nucleus clusters (ANCs) of the basic topology of chromosomes (nucleosomes, turns of nucleosomes in solenoid), as well as water and the studied mitochondrial DNA and NC1 domains. Heterogeneity was found in the bio field, which is expressed in extreme values of the number of nodes and antinodes, and the part of dense nodes. Signals from ANC allow remote monitoring of the state of their conformations in the basic topology of chromosomes, mitDNA, NC1 non-invasively, *in vivo* mode.

Keywords: bio field, egg, neutrino, halo, structure, mitDNA, nucleosomes, NC1

---

* Corresponding Author's Email: heide-lore@zubow.de.

In: Advances in Biology. Volume 4
Editor: Charles D. Grant
ISBN: 979-8-88697-999-2

Introduction

Previously, we discovered neutrino halos around various objects and described their isobars as super positions of standing neutrino waves [1]. It was of interest to understand the structure of neutrino halos around biological objects and the dynamics of the shock waves that make up the halo under various impacts on these objects.

The aim of this work was to study the neutrino halo around the chicken egg of a domestic hen, its structure and long-range action, as well as molecular contributions to the halo from the basic topology of chromosomes, mitochondrial DNA rings, and NC1 domains.

Material and Method

Unfertilized eggs of domestic chicken (*Gallus gallus domesticus*) were chosen as the object of study. The method of gravitational mass spectroscopy was used to study gravitational noise (GN) from oscillations of atomic nuclei clusters (ANC) as super positions of equilibrium proton processes with neutrinos, Figure 1, [2]. ANC – water clusters, domains, coils of proteins, DNA, nucleosomes, elements of the basic topology of chromosomes. Their oscillations generate a wide spectrum of GN. Gravity mass spectra (GMS) were obtained by subtracting the background GN from the GN of the sample. The value of Δf reflects the energy fraction of the ANC from the energy of the entire ANC ensemble in the studied GN frequency range. The ratio of the frequencies (ν, Hz) of shock waves (SWs) and the masses of ANC calculated for them (*m*, Dalton) was carried out using the first Zubow equation [2]. GMS spectra were obtained using this equation; the pressure of weak shock waves used in the spectrometer was less than 1 N/m^2; positive values of Δf are characteristic of loose clusters and domains; they reflect the energy fraction of the domain in the entire ensemble of studied masses, negative values of Δf reflect the same, but for dense coils. Scan time 10 s. N is the number of kinds of molecular clusters in the GMS spectrum (number of signals), $M_{GMS} = \sum(m \cdot |\Delta f|)$ is the average molecular mass of clusters in Daltons (Da), and D_c is the part of dense, potential energy rich clusters $(\sum(-\Delta f)/\sum(\Delta f)+(abs(-\Delta f))$. Figure 2 shows a schematic illustration of the method. The resulting ANC oscillations are determined by the equilibria of protons with neutrinos. It is accompanied by the generation of gravitational noise (GN) from the ANC. Hence, the GN is represented by a set of GN shock waves from an impulse shift in the equilibrium of freely oscillating protons (FOP, between dissolved in physical vacuum and baryonic forms) and baryonic protons, whose oscillations are loaded with absorption neutrinos, Figure 1.

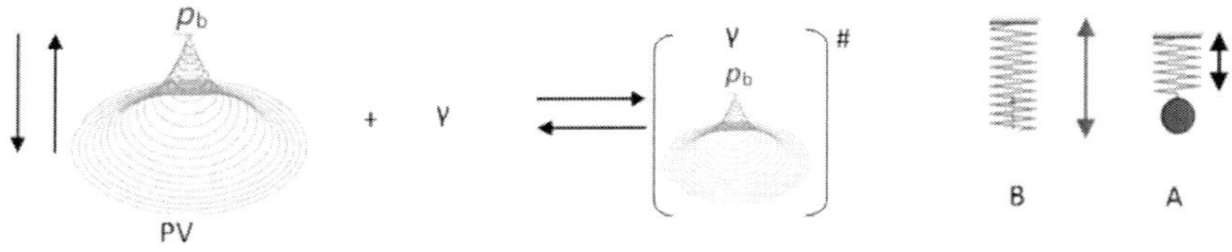

Figure 1. Model of proton oscillations (4.5E22 Hz, [2, 3]) between the physical vacuum and the baryon state, and the interaction of the proton with the neutrino. On the right, for understanding, a illustration of the "loading" of these neutrino proton oscillations is given. A - loaded, B - not loaded, states.

In the energy nodes of the GN (https://ru.wikipedia.org/wiki/Antinode, see also https://ru.wikipedia.org/wiki/Figures_Chladni) the states of the ANC with a minimum of potential energy are realized. This is where stable ANCs appear, that is, these states are resolved by the Universe. This is baryonic matter (https://ru.wikipedia.org/wiki/Baryonic_dark_matter). In the antinodes, on the contrary, ANCs are not resolved by the Universe, but they can temporarily appear as metastable states and then quickly collapse. At the same time, the GN of baryonic matter, for example, ANC, have a local effect on the energy of the nodes and antinodes of the field, which makes it possible to register ANC, calculate their masses, and get an idea of their conformations and the influence of the molecular surroundings. Therefore, the GMS method is convenient both for studying fields and states of ANC.

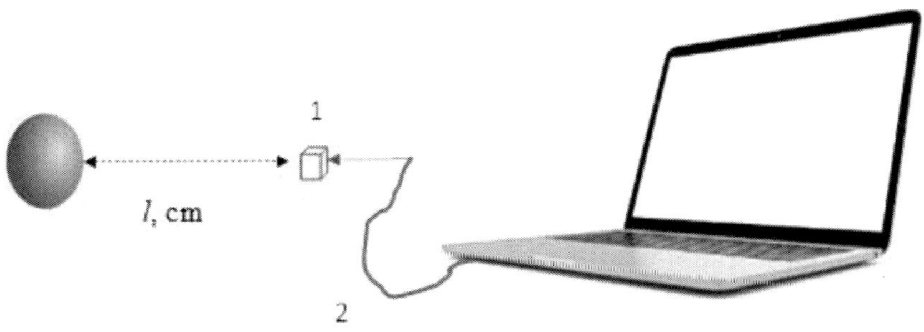

Figure 2. Schematic illustration of the setup for the study of the neutrino halo of the egg. 1 – GMS sensor, 2 – interface, *l* in cm – distance between the sensor and the egg shell.

To calibrate the GMS spectra, internal references were used, such as the basic water cluster $(H_2O)_{12}$ [4], the NC1 domain with a mass of 172,715 Da [5], the nucleosome octamer 132,982 Da [6], and also mitDNA 237,727,703 Da in sugar beet [7].

Results and Discussion

Figure 3 shows the states of the GN surrounded by a chicken egg. The number of energy nodes (N) and the parts of their densest states (D_c) at different distances from the egg were chosen as criteria. As can be seen from this Figure, the curves are not monotonic, they have a number of common extreme values. Anomalies of nodes at a distance of 15 cm stand out in particular. At more distant distances, the spread of values decreases, and the neutrino halo ends at 50 cm. Further, the sensor does not register the influence of the GN of the egg. Note that the neutrino halo is not stable in this range of offsets. It is influenced by the neutrino fields of the surroundings, for example, the presence of a human neutrino halo, our star, as well as the dynamics of the movement of massive clouds. But in general, extreme values persist. It can be seen that at night, the number of nodes in the field is less than in the daytime, and their contribution to the structure of the bio field is different. So at a distance of 15 cm, at night, the part of dense nodes decreases sharply, and in the daytime, on the contrary, increases. This means that in the daytime, the GN of the star is the energy source of the ANC eggs, which generate the bio field, and at night the planet takes away, absorbs the energy of the bio field of the ANC oscillator ensemble. A "hole" appears in the bio field at 15 cm.

In Figure 4, a model of a neutrino halo around an egg is given for understanding. The multi-layered nature of the halo indicates its complexity, as a superposition of the halo from a multitude of ANCs that make up the structure of the egg cytoskeleton. Let us try to understand the role of the ANC of the basic topology of chromosomes and large rings of the original (before genetic modifications of the bird) mitochondrial DNA in the neutrino halo of the egg.

It can be seen from the figure that both types of nucleosomes have similar curves. The sensor registers them as ANCs in dense conformations up to a distance of 50 cm. Then the halo blurs and its diffuseness manifests itself in the form of signal instability.

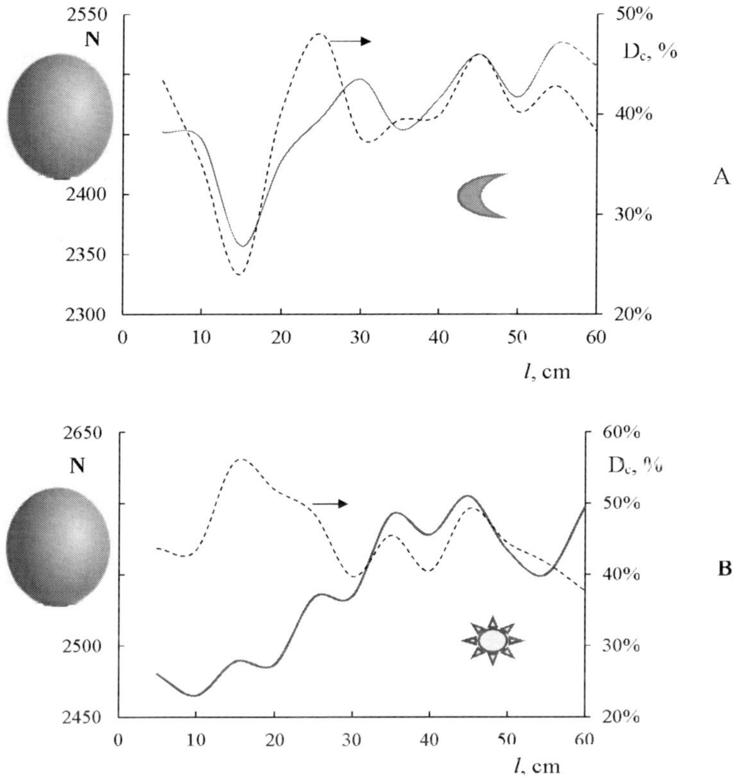

Figure 3. The number of nodes in the GN and the parts of their dense forms at different distances from the surface of an egg, 03/01/2023, 02:37 AM (**A** - night) and 03/02/2023 at 07:51AM (**B** - day) CET.

Figure 5 shows such results for octamers with different degrees of winding of DNA helices around them. Signals of only two types of nucleosomes were considered for analysis (see also Figure 10). One can see the presence in the bio field, mainly of nucleosome signals in dense conformations (-Δ*f*) up to the outer border of the halo. Recall that the energy nodes in the field are characterized by a minimum of potential energy, and the spread of energy values is minimal here. It is in them that the GN of the Universe allow the formation of nucleosomes. In contrast, in the antinodes, with fields with high potential energies, such ANCs are not allowed, the antinodes of the fields destroy them. However, in nature, metastable, transition states, states of unstable ANC are also possible. So, in the bio field, no signals of nucleosomes with one turn of the noose loop from DNA were found, but it appears in the boiled yolk (Figure 10**C**). This phenomenon should be understood as a forced transfer of ANC conformations, by the action of heat, into new metastable states. That is, thermal denaturation of macromolecules transfers them to states with high potential energy, and the GN of the Universe counteracts this. This means that a nucleosome with one DNA turn will need a large relaxation time to change conformation and time to rebuild the entire ANC ensemble, under the resonant field of its neighbors (https://www.defectologiya.pro/zhurnal/phenomen_bazovyix_molekulyarnyix_patologij_mozga_molekulyarnyij_mexanizm_formirovaniya_rechi_rebyonka_alczgejmer_i_parkinson/, and GN of Universe, and return in the potential field of the entire ensemble of oscillators in the long-range order of the egg with a minimum of potential energy, that is, to disappear.

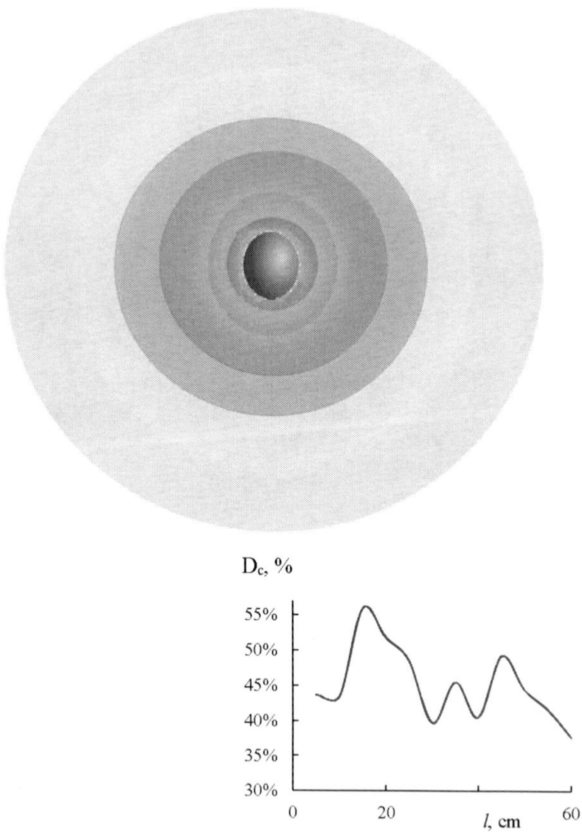

Figure 4. Structural model of a neutrino halo around a hen's egg at 07:51 AM, CET, 2.3.2023.

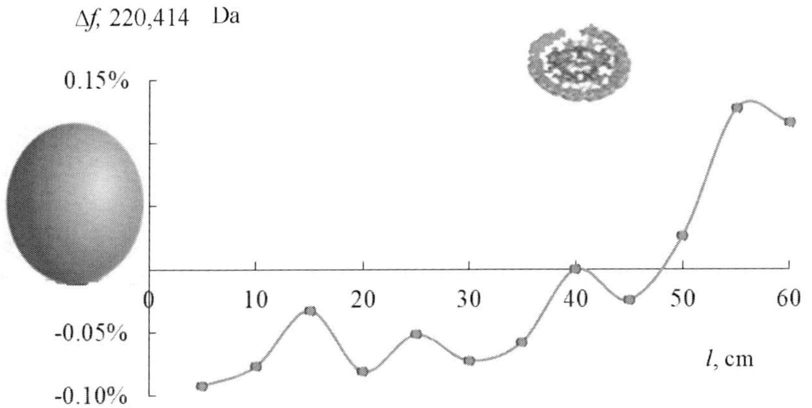

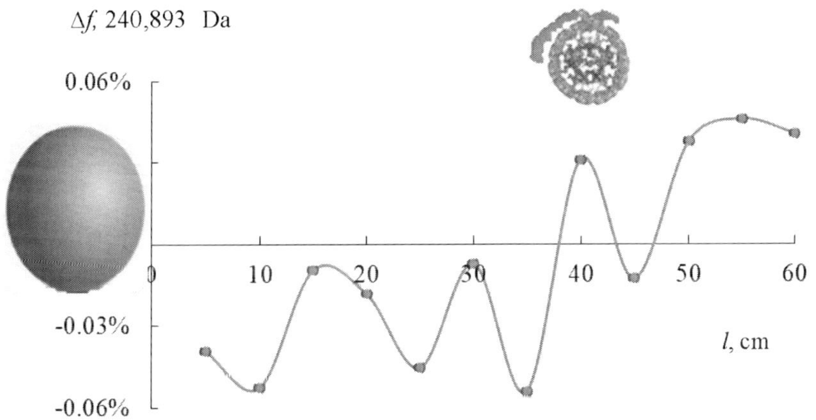

Figure 5. Contribution of neutrino halos of the basic topology of chromosomes to the superposition of the neutrino halo of the egg, Figure 4, 07:51 AM, CET, 2/3/2023. No signals were found from nucleosomes with one DNA strand. Here are the signals of the nucleosomes closest to this main nucleosome in mass, including the DNA link.

The contribution to the total neutrino halo from the oscillations of mitochondrial DNA large rings, which were preserved in the bird after numerous genetic manipulations, is shown in Figure 6. It can be seen from the Figure that at a distance of 15 cm there is an extreme value from the signals of mitDNA large rings (10,784,143 Da). Ring conformations are represented by loose structures with strong interactions with the surroundings. In the bio field, these are maxima at distances of 15, 30 and 40 cm.

Thus, ANC ensembles in the egg act as a united front of coherent oscillators in the formation of the neutrino halo of the egg. Such an egg has no pathologies (https://www.defectologiya.pro/zhurnal/fenomen_bazovyix_molekulyarnyix_patologij_mozga_molekulyarnyij_mexanizm_formirovaniya_rechi_rebyonka_alczgejmer_i_parkinson/).

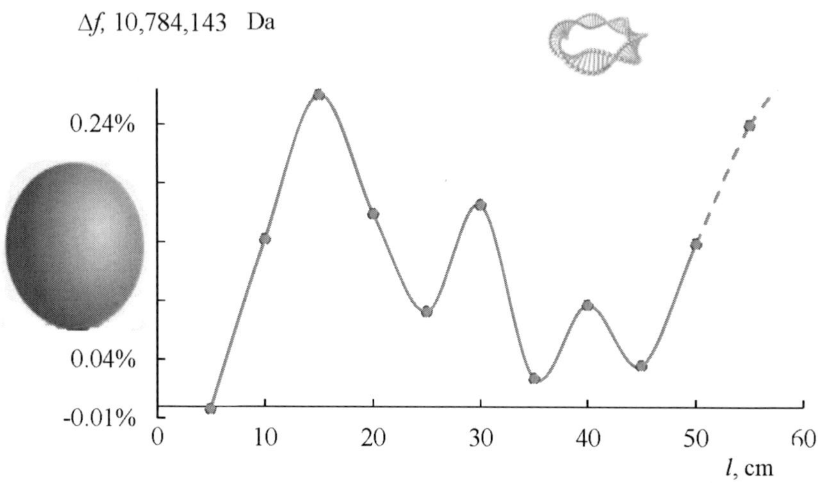

Figure 6. Contribution of neutrino halos from oscillations of large mitDNA rings to the superposition of the egg neutrino halo, Figure 4, 07:51 AM, CET, 2/3/2023. Commentary in the text.

Let us briefly stay on the contribution to the egg bio field of oscillations of other elements of the basic topology of chromosomes, for example, oscillations of turns of solenoids built from 6 and 12 nucleosomes [8, 9], Figure 7. Here, up to distances of 25 cm, significant differences are found. If the solenoid structures built from coils consisting of 12 nucleosomes fit satisfactorily into the general field of the neutrino halo, then the solenoid structures with coils of 6 nucleosomes are bad. Apparently, this is due to the strained degree of turn packing in the solenoids and the short DNA link between nucleosomes in one turn. In Figure 7, the package is shown as a dense bunch of turns in isotactic configurations.

It was of interest to understand the state of the bio field when the egg was kept at 310 K. Figures 8 and 9 show the curves of changes in the number of nodes and the part of their dense forms in the GN of the egg bio field. Comparison of the data of both Figures leads us to the understanding that the incubation time affects the state of the bio field due to changes in the GN itself.

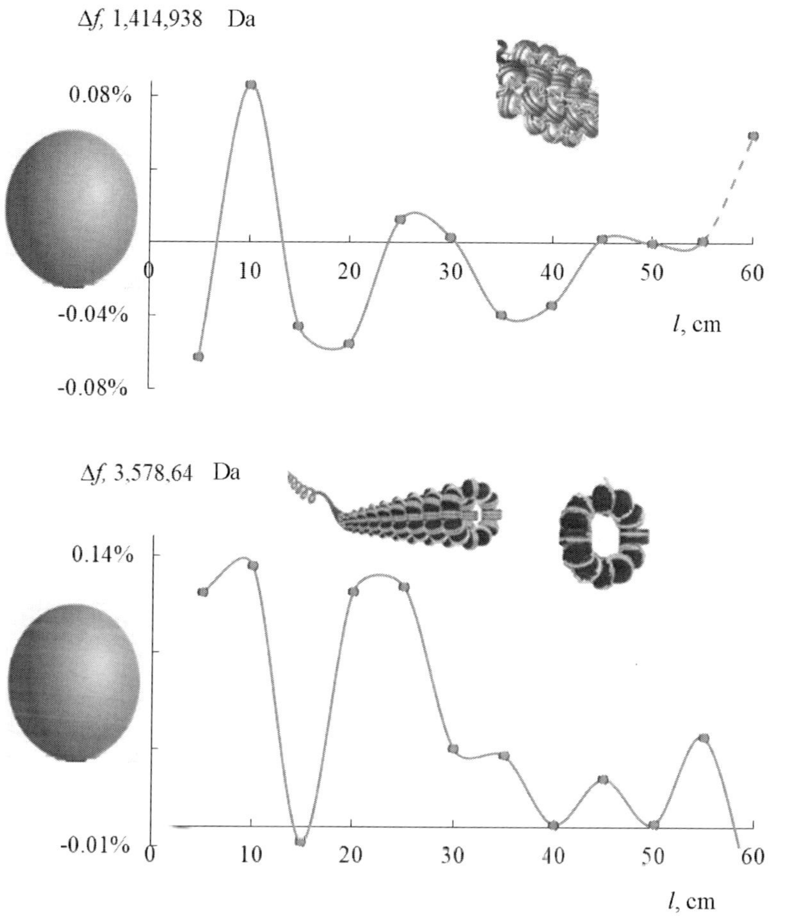

Figure 7. The contribution of the oscillations of the turns of solenoids, like ANC, to the basic topology of chromosomes, consisting of 6 and 12 nucleosomes in each turn in the bio field of the egg. The signals of turns were taken, like ANC, built from nucleosomes with a mass of 240,893 Da, including the DNA link (Figure 5).

Thus, the number of active ANCs in the GN with an initial spread of ΔN ~ 150 increased by 48 hours to a spread of values of ΔN ~ 250, and the part of dense energy nodes and, consequently, ANCs in them, sharply decreased, with a spread of D_c ~ 20% (start) to a scatter of ~5% (48 h). What should be understood as the harmonization of the entire ANC ensemble with thermal GNs and the appearance of their new thermodynamic state, minimization of the potential energy of the entire ANC ensemble, and, further, ordering of the long-range order in the egg cytoskeleton.

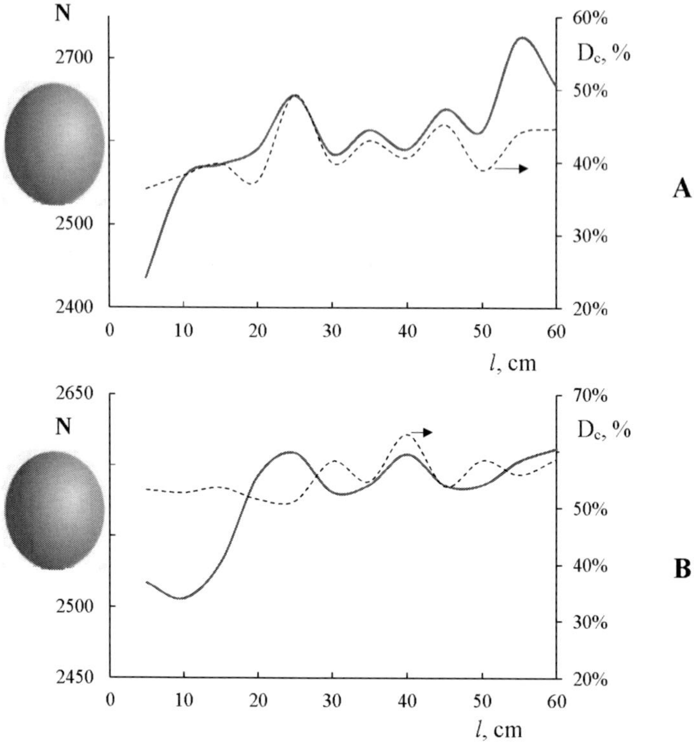

Figure 8. The number of nodes (N) and the part of their dense forms (D_c) in the bio field of the egg at 310 K. **A** - 48 h (5/3/23 at 07:59 AM, CET), **B** - 6 h (3/3/23 at 07:26 AM, CET) and 0 h, see Figure 3B.

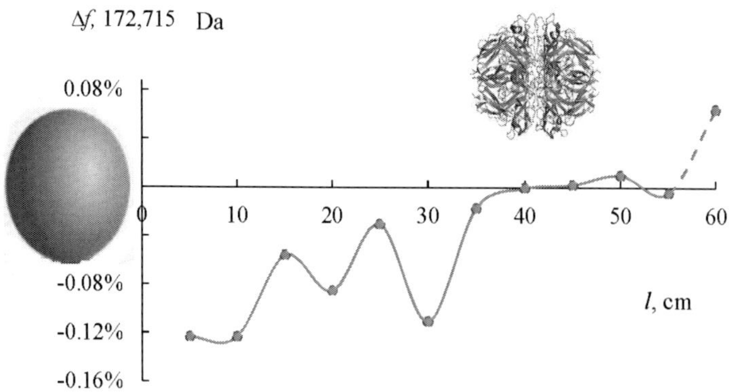

Figure 9. GMS signals of NC1 hexamer oscillations (nodes in the GN at 760 ± 10 Hz) at different distances from the egg, 07:51 AM, CET, 2/3/2023, 310 K.

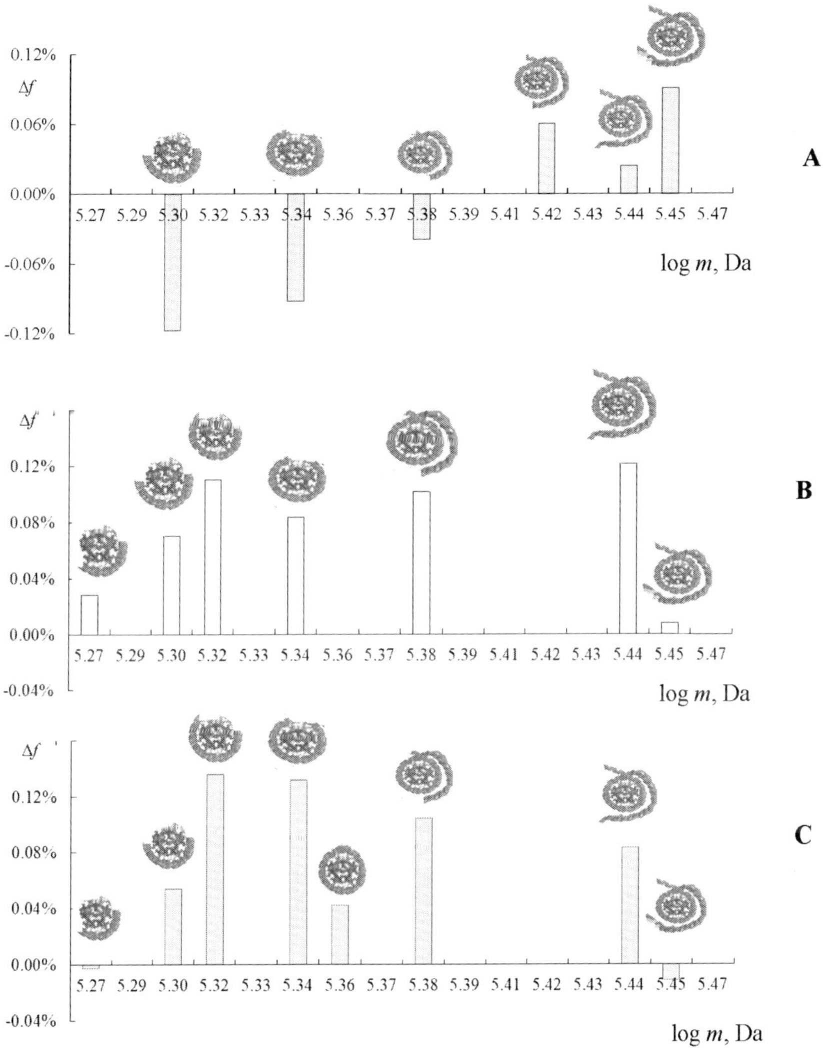

Figure 10. The state of nucleosome conformations in the egg. **A** - spectrum of the biofield in 5 cm from the surface of a fresh egg and in this egg after 4 minutes of boiling at 373 K, **B** - in the mass of boiled protein and **C** - in the mass of boiled yolk.

The state of GN, which is responsible for the formation of the terminal NC1 hexamer in collagen, is shown in Figure 9. It also shows the heterogeneity of halo with minima at <10, 20, and 30 cm, in which this domain should manifest itself as ANC in dense conformations, i.e., a domain with a minimum of interaction with its surroundings. In comparison with Figure 5, for GN that initiate the appearance of other ANCs (nucleosomes), one can notice a common feature only in the immediate vicinity of the egg. At longer distances, the spectra differ.

The influence of strong thermal noise at 373 K on halo nodes where nucleosome formation is allowed (658 ± 20 Hz) is shown in Figure 10. It can be seen that the GN signals from nucleosomes in the fresh egg halo (**A**) differ from those directly in the mass of boiled proteins (B) and yolk (**C**). At the same time, nucleosome signals with one turn of the DNA strangle helix appear in the yolk.

In the region of oscillations that are not completely covered with DNA of octamers, the number of their species is less in the GN halo **A**, in 5 cm from the shell of a fresh egg, than directly in the mass of boiled white and yolk. They also differ in the degree of interaction with the surroundings. Here we come to the conclusion that a strong thermal effect at 373 K not only increases the number of types of incomplete nucleosomal constructions, but also increases their degree of interaction with the surroundings, i.e., loosens out of conformation.

Conclusion

Chicken eggs have a neutrino halo - a bio field.

The neutrino halo propagates within a radius of up to 50 cm. The maximum heterogeneity in the bio field takes place within a radius of 15 cm.

The bio field is a superposition of bio fields from ANCs in the biomolecules of yolk and protein.

The GMS method makes it possible to carry out a remote non-invasive analysis of the state of the basic topology chromosomes of cell and large rings of the initial (before genetic manipulations) mitochondrial DNA of the egg, NC1 hexamers and water *in vivo* mode.

Apparently, a chicken is able to recognize its eggs not only by size, color, but also by bio field.

ANCs in a chicken egg are formed in the gravitational noise of the Universe.

Their number and forms can be transferred to other, but metastable states by local GN, for example, GN of thermal effects.

References

[1] Zubow A., Zubow K., Zubow V. A. Phenomenon of Axions' and Neutrinos' Clouds around Objects. Nature of inertia. Editor A. Reimer. *Horizons in World Physics*. NY. 2014, vol. 282, pp.53-70. https://www.novapublishers.com/catalog/product_info.php?products_id=50340&osCsid=994455a7afd05deb707594aaf444ed06. *International Journal of Condensed Matter, Advanced Materials, and Superconductivity Research*, 2014, volume 13 Issue 1, pp. https://www.Novapublishers.com/catalog/product_info.php?products_id=53555&osCsid=9dc3348f7b2274186a8949c752b7f142.

[2] Zubow K., Zubow A. V., Zubow V. A. The Way to the ETIs. *Applied gravitational mass spectroscopy*. Nova Sci. Publ. NY, 2014. https://www.novapublishers.com/catalog/product_info.php?products_id=42668&osCsid=5bd85d42dc273360fd48126de7bc9daf. https://www.novapublishers.com/catalog/advanced_search_result. Php?keywords=Zubow%20K&sort=2a&page=3&osCsId=59237f459eafe861a2db593a6a504199.

[3] Zubow K., Zubow A. V., Zubow V. A. The Phenomenon of Proton Dissolving in Vacuum and of Proton Condensation from Vacuum. Two Forms of Protons, Structure of Nuclei, Electrons and Atoms. *J. of modern physics*, vol.1, no.1, 2010 pp.175-184. https://www.researchgate.net/publication/228928173_The_ Phenomenon_of_Proton_Dissolving_in_Vacuum_and_of_Proton_Condensation_from_Vacuum_Two_Forms_of_Protons_Structure_of_Nuclei_Electrons_and_Atoms.

[4] Annika Lenz, Lars Ojamäe. On the stability of dence versus cage-shaped water clusters: Quantum-chemical investigations of zero-point energies, free energies, basis-set effects and IR spectra of (H2O)12 and (H2O)20. *Chem. Phys. Letters*. 2006, v. 418, pp. 361-367.

[5] Manuel E. Than, Stefan Henrich, Robert Huber, Albert Ries. The 1.9- α crystal structure of the noncollagenous (NC1) domain of human placenta collagen IV shows stabilization via a novel type of covalent Met-Lys cross-link. *Proc Natl Acad Sci U S A*. 2002 May 14, vol. 99, no.10, pp. 6607–6612. doi: 10.1073/pnas.062183499.

[6] https://ru.wikipedia.org/wiki Нуклеосома.

[7] Tomochito Kubo, Satsuki Nishizawa, Akira Sugawa, Noriko Itchoda, Amy Estiati and Tetsuo Mikami. The complete nucleotide sequence oft he mitochndrial genome of sugar beet (Beta Vulgaris L.) reveals a novel gene for tRNA cys(GCA). *Nucleic Acids Research*, 2000, Vol.28. No. 13, pp. 2571-2576.

[8] Henikoff S, Shilatifard A. Histone modification: cause or cog? *Trends in Genetics*, October 2011, Vol. 27, No. 10, pp. 389-396, doi:10.1016/j.tig.2011.06.006. https://research.fhcrc.org/content/dam/stripe/henikoff/files/2011%20histone%20modifications.pdf.

[9] Diana Lang. *Proteomische Methoden zur Identifizierung acetylierungsabhängiger Interaktionspartner von Histon H4* [*Proteomic methods to identify acetylation-dependent interaction partners of histone H4*]. Dissertation von der Fakultät II – Mathematik und Naturwissenschaften der Technischen Universität Berlin, 2012. https://depositonce.tu-berlin.de/bitstream/11303/3575/1/Dokument_ 47.pdf.

Contents of Earlier Volumes

Advances in Marine Biology. Volume 3

Advances in Biology. Volume 2

Advances in Biology. Volume 1

Index